T0074484

ATOMIC ENVIRONMENTS

NE**X**US

NEW HISTORIES OF SCIENCE, TECHNOLOGY, THE ENVIRONMENT, AGRICULTURE & MEDICINE

NEXUS is a book series devoted to the publication of high-quality scholarship in the history of the sciences and allied fields. Its broad reach encompasses science, technology, the environment, agriculture, and medicine, but also includes intersections with other types of knowledge, such as music, urban planning, or educational policy. Its essential concern is with the interface of nature and culture, broadly conceived, and it embraces an emerging intellectual constellation of new syntheses, methods, and approaches in the study of people and nature through time.

ATOMIC ENVIRONMENTS

Nuclear Technologies, the Natural World, and Policymaking,
1945–1960

✳——————————————————————————————✳

NEIL S. OATSVALL

THE UNIVERSITY OF ALABAMA PRESS   TUSCALOOSA

The University of Alabama Press
Tuscaloosa, Alabama 35487–0380
uapress.ua.edu

Typeface: Scala Pro

Cover image: Operation Crossroads, Bikini Atoll, July 25, 1946;
United States Department of Energy
Cover design: Michele Myatt Quinn

Cataloging-in-Publication data is available from the Library of
Congress.
ISBN: 978-0-8173-2146-8
E-ISBN: 978-0-8173-9434-9

For my dearest loves, Eleanor and Abigail

Contents

Illustrations

Foreword

In 1977, the eminent environmental historian Donald Worster noted that "The Age of Ecology opened on the New Mexican desert, near the town of Alamagordo, on July 16, 1945, with a dazzling fireball of light and a swelling mushroom cloud of radioactive gases." In Worster's telling, the appearance of a technological force that could destroy life on Earth led to a collective reckoning about how people ought to wield that power. In the years since, historians have expanded and complicated that understanding as they have established the links between environmental science and atomic energy. Myriad studies now explore various facets of the connections between nuclear and environmental science, taking up subjects ranging from the ways in which nuclear science proved fundamental to the development of sciences such as ecology to efforts to use atomic energy to improve agriculture, from the role of the military in shaping Cold War science to the place of nuclear technology in efforts to weaponize the natural world itself.

In this important new book, Neil Oatsvall draws on deep archival research to augment these studies by shifting the focus to the high-level, civilian decisionmakers at the Atomic Energy Commission (AEC). Telling the story from their perspective, Oatsvall uncovers the ways in which they valued environmental knowledge, not for its own sake but because it allowed them to direct nuclear policy and investment. In doing so, Oatsvall offers a salient reminder that not everyone who thought deeply about the natural world did so through an environmentalist lens. Moreover, he inverts a more common emphasis on the ways in which atomic science shaped the development of environmental science. Through the eyes of AEC policymakers, it was clear that the development of nuclear technology hinged on the development of environmental science no less the reverse. Indeed, the two were of a piece.

In Oatsvall's telling, then, it is not Alamogordo, or Project Plowshare, or the tracing of radioisotopes through plants, or military funding that typified the intersection of environmental and nuclear science in Cold War America but the Encore test shot of 1953, which took place over one hundred-forty-five mature ponderosa trees cemented to the desert floor so that the effects of a nuclear blast on a forest could be studied. The manifold ironies of such events are not lost on Oatsvall, but they prove less important to him than the seriousness with which they were undertaken.

His analysis begins with a focus on the ways in which nuclear testing forced a consideration of environmental factors—in selecting testing sites, in wrestling with fallout, and in detecting covert nuclear testing. From there it shifts to an examination of the ways in which AEC decisionmakers sought to win public support by leveraging the potential benefits of an "atomic agriculture." Those motives matter, but so too does the fact that those efforts demanded that policymakers draw upon environmental science to frame a persuasive case. He closes with a similar case, highlighting how an understanding of the functioning of local environmental systems proved essential to finding an appropriate place to discard nuclear waste.

Oatsvall offers, in many ways, a depressing story. In the end, after all, the policymakers repeatedly made decisions that later generations would rue. Too often, they privileged public relations or institutional concerns over environmental and human health. It is, thus, to Oatsvall's credit that he tells the story in a lively and engaging way. And if other historians might judge the AEC's decisionmakers more harshly than Oatsvall, by taking AEC officials seriously on their own terms, Oatsvall has rendered it impossible for future historians to imagine them indifferent to the natural world. Moreover, he has unearthed new opportunities for historians to explore the ways in which other groups—from the courts to ordinary citizens—shaped the intersection of nuclear and environmental science during the first decades of the Cold War. For if Oatsvall's AEC commissioners often had the president's ear, they were not the only ones; they were hardly alone in shaping nuclear policy.

By design Oatsvall's study isn't the final word on the role of policymaking in the shaping of environmental and nuclear science. But it is a study that will need to be reckoned with going forward. Indeed, in opening space for

future studies at the intersection of environmental history and the histories of science and technology, this book fits well within the aims of the NEXUS series, and we warmly welcome its publication as part of the NEXUS series.
—MARK D. HERSEY ON BEHALF OF THE NEXUS EDITORS

Acknowledgments

I knew I would incur many debts while researching, writing, and publishing this book, but I had no idea those would be so numerous and deep. Though the small thanks I add here do little to repay those listed below, my gratitude is significant.

Professionally, so many folks provided selfless encouragement, advice, mentorship, and close readings of the manuscript in whole or part at various stages. My thanks to Sheyda Jahanbani, Greg Cushman, Sara Gregg, Johan Feddema, Ted Wilson, and Don Worster for the assistance at early stages of the manuscript. Thanks also to archivists at the Dwight Eisenhower Presidential Library, the Harry S. Truman Presidential Library, and the National Archives in Washington, DC. Heidi Palombo at the Department of Energy was especially helpful in securing many of the images in this book. Other friends and colleagues provided great help with the manuscript at later stages over the years, including Lisa Brady, Karl Brooks, Jake Hamblin, John Hess, Josh Nygren, Adam Rome, Vaughn Scribner, and Julia Adeney Thomas. This book would have been much worse without their insight, commentary, and suggestions.

I would be remiss without giving a special thanks to Mark Hersey, a long-time friend who recruited me to the NEXUS series and provided more help in this book project than anyone else. It has been a professional and personal joy to add "book editor" to our relationship, and many of the best parts of this book are owed to him. Alan Marcus and Alexandra Hui rounded out the series editorial team and provided useful feedback as well. On the press side, thanks to Beth Motherwell for starting me through the publication process at

University of Alabama Press and especially Claire Lewis Evans for shepherding the manuscript through the trickiest stages of the publication process. And, though I may have cursed you several times over the years, my sincere and eternal gratitude to the three anonymous reviewers who participated in two rounds of peer review for this book while still a manuscript. Your suggestions and critiques made this book much stronger than my initial submission. (And to anyone else out there who gets a negative peer review, do not get discouraged! It happens to everyone at some point.)

My previous institution, the Arkansas School for Mathematics, Sciences, and the Arts, provided not only a solid home base during much of this project but, more important, gave me a partial course release at times to work on this research and other projects. Thanks to my many colleagues and friends for their support, companionship, and encouragement, including Bryan Adams, Tom Dempster, Brian Isbell, Stuart Flynn, James Katowich, Dan Kostopulos, Mary Leigh, and Dan McElderry. Some people are missing intentionally from this list, and others it was probably unintentional. My apologies if you are in the latter category.

University of Washington Press published a portion of chapter one in *Proving Grounds: Weapons Testing, Militarized Landscapes, and the Environmental Impact of American Empire*, edited by Ed Martini. The journal *Agricultural History* published a portion of chapter four. Thanks to the editors and anonymous reviewers involved in those publications for their time and sharp comments.

On a personal level, there is little I can say that would indicate the depths to which I have depended upon friends and family over the years. Bill Velto has been an advocate for me and my research for longer than I knew I needed it. My parents, Mike and Becky, and my sister, Jessi, and her family Michael, Mackenzi, and Logan, have given me unconditional love, and it is reciprocated. Other thanks to Allen and Mary Lee, who moved to Arkansas in the middle of this project, and also Cecelia, Jason, Bo, Bebe, Lee, Martha, Alec, and Sallie, may she rest in peace.

Of course my greatest thanks are reserved for Sarah, who has provided more support and encouragement than anyone, and our daughters, Eleanor and Abigail. Nothing I write here will do justice to the depth of love and adoration I have for the family and life we have built together, no matter the form that takes. As always, all of it is for you three.

Finally, it probably does not need to be said, but I will anyway. Any mistakes that remain in this book are mine and mine alone, even if there are plenty of folks I would happily blame if given the opportunity.

ATOMIC ENVIRONMENTS

Introduction

During the spring of 1953, the United States Forest Service began a construction project in Nevada. Forest Service personnel were not clearing a campsite or hammering together benches, as they might have been expected to do during this boom period of national parks. Instead, they created a forest, of sorts. Workers selected 145 mature ponderosa pines from nearby "forest reserves," harvested those trees, and transported them to a grove around 150 feet wide and 300 feet long. Once there, Forest Service personnel used cement to secure the pines into their best approximation of a natural forest environment.[1] In most woodland ecosystems, lush, verdant growth offers refuge and sustenance to many animals, birds, and insects. This forest was, by contrast, devoid of life. Into this sterile landscape, workers from the Atomic Energy Commission (AEC) scattered instrumentation about, effectively transforming the woods into an outdoor laboratory. Relying upon the knowledge and expertise of the Forest Service, the AEC saw the finished coppice as an idealized forest. But the agencies never intended their new forest to flourish in any traditional ecological sense. To the contrary, the Forest Service and AEC created the forest as a test laboratory that could be destroyed in a controlled and measurable fashion. In May 1953, the AEC carried out the Encore test shot of the Operation Upshot-Knothole test series and detonated an atomic bomb near the anthropogenic forest.

Analyzing the forest's demolition within a Cold War context, the AEC hoped to determine how much an atomic weapon damaged a stand of coniferous trees and in turn assess how much protection (or extra damage from shrapnel) the forest might provide nearby persons and materials.[2] Creating an idealized forest, in this case, was actually more useful than a more natural space might have been because it allowed testers to isolate variables and gather data more easily. Hence while the AEC did not expect the forest to thrive ecologically, decision makers there did care a great deal about the forest for what it could tell them about the nation's nuclear arsenal. Ecological knowledge, in this case, buttressed nuclear science. Planners hoped that the test would improve the nation's understanding of how best to protect US resources and personnel during a nuclear war.[3] Destroying the constructed forest arose, ironically, from an impulse to protect other US forests and the humans near them from harm.

The Encore test itself created an impressive sight when it went off a mile or so from the newly manufactured forest. Initial heat from the nuclear detonation started a conflagration. The successive blast annihilated much of what remained. All told, the trunks, branches, and needles were severely damaged, which can be seen from a video recording of the test.[4] How the blast would have affected root structures is indeterminable since those were surely mangled when the Forest Service uprooted the trees from their original location and cemented them back into the earth.

The Encore test shot represents more than a fascinating spectacle or a curious footnote in the history of US nuclear projects. At first glance, the test fits neatly into a common paradigm based on assumptions that nuclear weapons (and other nuclear technologies after the disasters at Chernobyl and Fukushima) can only have harmful interactions with world environments. Such a perspective envisions nuclear technologies as the pale horse rider of the Apocalypse, charging forth and spreading death.[5] For example, in 1968, Sheldon Novick, Barry Commoner's research assistant, skewered the AEC for what he perceived as the reckless production of nuclear wastes via nuclear power production. In "The Menace of the Peaceful Atom," Novick proclaimed, "Once they are released into the atmosphere, there is no conceivable way of retrieving radioactive gases; once entered on their winding course through the environment, radioactive isotopes are out of reach of man's control. The damage, once done, is irremediable. If we do not begin now to invest the ingenuity and the money which are necessary to prevent the release of radiation through the commercial use of atomic power, we will end by damaging the very fabric of life on this planet."[6] More recently, antinuclear advocate Helen Caldicott summed up the thoughts of many when she belted out at a 2011 rally, "The nuclear industry is a death industry. It's a cancer industry. It's a bomb industry. It's killing people and will for the rest of time."[7] Such an antagonistic approach might smoothly transition into accusations that the AEC nuked a forest for America.

Yet in spite of the Encore test being designed with explicit destructive intent, planners both depended on and sought to promulgate environmental sciences with the experiment. Because one of the test's stated goals was to learn how better to protect US forests from nuclear attacks, planners constructed the experiment to safeguard and not destroy forests. If protecting people was paramount in the AEC's planning, the natural world provided more than just a bomb target for the US nuclear program—it offered a new vehicle for knowledge about nuclear technologies and a way to understand what those technologies meant for the safety and wellbeing of the United States. To AEC testers, the moment when their fabricated forestland met that nuclear blast represented the culmination of intersecting and interdependent intellectual currents in nuclear and environmental science.

The Encore test clearly demonstrates how environmental considerations impacted the development of the US nuclear program. Historian Ferenc M. Szasz argued that scientists turned the location of the first atomic bomb detonation, the Trinity site at Los Alamos, into "a sprawling, open-air scientific laboratory" that melded science and the natural world into one indistinguishable mass.[8] Yet the Encore test went beyond even Szasz's assertions. The nuclear blast and the newly built forest never existed independently, ecologically, or in planners' minds. The forest allowed testers to build a more effective bomb and perhaps better protect troops from nuclear weapons. And, springing forth from fears of a nuclear-environmental disaster, the weapon helped testers understand how woodlands function when engulfed in a nuclear holocaust. The Encore test shot rested upon twin pillars of nuclear and environmental science, revealing that early Cold War policymakers needed to understand and implement scientific knowledge about the natural world to advance nuclear technologies, depending on nuclear and environmental science each to support the other.[9] This book brings policymaking about the juncture of environmental and nuclear science to the forefront.

Atomic Environments employs two broad, interrelated goals to help place the interconnections between nuclear technologies and the environment into their practical applications. First, the book seeks to uncover how policymakers questioned the extents and limits of their responsibilities to the nation, its government, and its peoples. In some ways, this idea gets at the heart of what constitutes politics. But in other ways, it is a fundamental recognition that the responsibilities these decision makers perceived themselves as having could at times conflict with each other. While policymakers intended various nuclear technologies to safeguard national interests or improve their constituents' way of life in some capacity, every nuclear technology also interacted with the natural world, sometimes to the detriment or at least potential detriment of the very peoples and places policymakers were supposed to protect. Second, these high-level administrators grappled with how to assess the limits and capabilities of nuclear technologies and how those limits ought to be managed. In doing so, they encountered vexing questions about their role in the federal government and US society writ large. The studied policymakers clearly felt, in the geopolitical context of their time, that they could keep the country and its citizens safe from harm by developing, maintaining, and improving a significant nuclear arsenal. But whether the benefits of various nuclear technologies outweighed the drawbacks is difficult to pin down in hindsight and was especially troublesome for policymakers to confront at the time.

These two goals are, in many ways, about uncovering how decision makers, when confronted with new technologies and all the uncertainties involved in developing and implementing such innovations, utilized existing intellectual networks to make policy. The answer clearly involved interpreting scientific understandings about nuclear technologies and the environment as best they could, but a great many conundrums proved difficult to solve. This book argues that policymakers confronted these crises by turning to environmental science and basing their decisions on environmental knowledge more often than has generally been acknowledged. Exploring the ways in which they did so can tell us a great deal about the interplay between an expansive nuclear culture in the postwar United States and the natural world.

The fundamental argument of *Atomic Environments* is therefore that developing nuclear technologies within an early Cold War geopolitical context necessitated that US policymakers utilize and foster environmental science throughout their decision-making. The natural world and the scientific disciplines that study it, instead of holding back nuclear technologies, became integral components of nuclear science. In this way, atomic science was not merely a despoiler of world environments, affecting humans' relationship to the natural world only as an ironic, unintended impulse for the coalescence of the modern environmental movement. Those same technologies also functioned as a vehicle for improving scientific understandings of the environment and involving those into policymaking. Taking seriously the overlapping research agendas (indeed the coevolution) of nuclear and environmental sciences offers historians a fresh look at nuclear technologies, nuclear research and development, and the policymaking that facilitated both.

An ecological approach to nuclear technologies therefore amplifies the work of those historians like Joel Hagen who have shown that nuclear technologies were fundamental to the development of environmental science.[10] This book demonstrates that the reverse is also true—nuclear technologies developed in partnership with various environmental sciences.

The story that emerges, however, is less about individual policymakers than the institutions in which those policymakers served. No matter who made the decisions, public relations and nuclear boosterism were consistently privileged over frank assessments of the risks and dangers of the nuclear age. While individuals in the US government may have held genuine concern for how the development of nuclear technologies might harm the United States and its peoples, such an institutional position rarely emerged. While this book is not quite a collective biography of the involved decision makers, it is indeed more focused on the institutions that held a fairly constant set of values over the first two nuclear age presidencies spanning more than a decade and a half than it is any individual person. But an institutional lack of concern for environmental wellbeing did not equate to lack of attention paid to the natural world. When historians Mark D. Merlin and Ricardo M. Gonzalez cataloged the "direct and indirect atmospheric, geological, and ecological effects of nuclear testing in Remote Oceania," they claimed, "most, if not all, [US tests in the Pacific Ocean] were initiated with explicit political intention, often with little regard for the ecological consequences."[11] But such assertions run contrary to the utilization of environmental science, especially meteorology, geology, ecology, and biology, in order to understand the areas in which tests occurred. US policymakers certainly cared about how tests might affect local and global ecology and used ecological damage to understand better the weapons they were testing—those policymakers just did not prioritize environmental or ecological health in their decision making. Instead, the situation is much closer to what Mark Fiege described when studying the Manhattan Project: "the nation's atomic project, especially the bomb, was deeply embedded in the human relationship to nature."[12]

Highlighting the connection between policymaking and the environment therefore allows *Atomic Environments* to give a nuanced examination of environmentally influenced policymaking, showing what early Cold War policymaking looked like when leaders took seriously both the natural world and lessons that could be learned from it. Historian Adam Rome has described the advent of the "environmental management state," the study of which, as historian Paul Sutter clarified, caused American environmental history to move "away from whiggish histories of the rise of an environmentalist sensibility and toward explorations of the varieties of environmental knowledge."[13] Unsurprisingly, few of the decision makers in this book could rightly be considered environmentalists in any meaningful way. But neglecting environmental considerations that were not explicitly environmentalist distorts our understanding of the historical record. Nuclear technologies heavily influenced the culture, politics, and institutions of the postwar nation state, and the environment figured heavily in their development. By extension, environmental knowledge figured crucially in the development of the postwar United States.[14]

While it takes more than knowledge of ecology and a desire to implement its findings into policy to engender environmentalist sentiment, recent scholarship has clearly demonstrated that policymakers did possess a significant amount of environmental and ecological knowledge. Jacob Darwin Hamblin's *Arming Mother Nature* showed how postwar politicians, bureaucrats, and military minds sought to harness the power of the natural world, such as through a natural disaster, for destructive purposes. Why bomb a city if you could create a colossal earthquake and unleash that on an enemy?[15] The framing in Hamblin's influential study underscored the ways in which interactions between scientific researchers and military officials influenced the production of scientific knowledge and its implementation into policy. This book takes a slightly different tack, studying not only the inclusion of environmental science into the US nuclear program but also how politicians, bureaucrats, and their institutions mediated those understandings. It is about how high-level executive policymakers valued environmental knowledge for how it could help them oversee the creation of a variety of nuclear technologies, some of which could even threaten the ecological health of the entire planet. *Atomic Environments* is therefore less of a history of science than a history of how environmental ideas became essential to the institutions of nuclear policymaking at the highest levels.

Since language can be imprecise and terms fairly malleable, brief clarification on the verbiage this book uses is worthwhile. The book uses *nuclear technologies* in a way inclusive not just of nuclear bombs or nuclear power reactors but also radioisotopes, efforts at nuclear waste disposal, and varied other applications. *Environment* and *natural world* are used here in the vernacular sense, roughly meaning the nonhuman natural forces and features with which nuclear technologies and policymakers interacted, physically and mentally. While scholars may debate the exact intellectual limits of such ideas, such as historian Etienne Benson in his fine book *Surroundings*, the people this book studies exhibited little such ambivalence.[16]

Perhaps trickiest is defining what is meant by *policymaking*, and *policymakers* or *decision makers*. The people and processes labeled as such and studied in this book represent an admittedly very narrow component of US nuclear politics. While modern political scientists most commonly study "issue networks," or the loose and somewhat informal connections existing between diverse groups and actors working in broad policy areas, nuclear policymaking at the time better fits an older idea of an "iron triangle" of policymaking. Iron triangles described somewhat "ironclad" relationships and interactions between three different sides of policymaking—bureaucratic agencies, congressional committees or subcommittees, and interest groups.[17]

When *Atomic Environments* studies *policymakers*, it is mostly only studying one side of that iron triangle—the executive, (largely civilian) bureaucratic side. Yet that set of actors is a bit difficult to tease out of the historical narrative because the US military, part of the executive branch and firmly enmeshed into the AEC, effectively served as the interest groups within that "iron triangle." While historian Brian Balogh well demonstrated that the iron triangle broke down over time in nuclear power because no effective interest groups ever formed, the problem in this story is that effectively one group occupied two sides of the triangle.[18]

Despite the focus of this book, the sum set of persons who actually made nuclear policy was incredibly broad during the time period. For example, both the military and the Congressional Joint Committee on Atomic Energy exercised tremendous influence in charting the nation's path toward nuclear weapons, nuclear power, and a range of other technological applications. For that matter, as historian Karl Brooks has argued, members of the judicial branch, ordinary citizens, and even the environment itself have also proven important in shaping policy and law.[19] But it was at the highest reaches of decision making where the synthesis of environmental and nuclear policymaking was most evident. While scientists and lower-level bureaucrats certainly engaged in both realms from time to time, studying US policymaking at the very highest levels is most valuable when trying to glimpse the crossover and melding of environment and atom.

This book therefore limits itself to high-level members of the executive branch, most frequently AEC commissioners (the commission's top level administrators) and officials in the White House, especially presidents Truman and Eisenhower. Employing such a precise definition of policymaking helps create a robust sense of how a particular set of policymakers—those at the very top concerned with the US nuclear program on more of a conceptual level—conceived of US nuclear technologies and depended on the natural world to help understand how those new technologies functioned within real-world contexts. Moreover, such a definition allows for examination of broader institutional positions and priorities that cut across individual decision makers. No matter those institutional positions, nuclear technologies could indeed pose a significant threat to humans and their environments.

Public imagination was actually often ahead of scholars (and policymakers) in terms of reflecting upon the effects radiation might have on biological entities. They read about the real effects of nuclear weapons in John Hersey's *Hiroshima* (1946) and saw Hollywood movies like *Them!* (1954), which portrayed a nightmare scenario of how long-term radiation exposure might mutate harmless ants into terrible monsters the size of Volkswagens.[20] While academics have produced a voluminous body of literature on nuclear technologies in the intervening years, the historical literature has generally avoided the natural world and instead focused on other aspects of nuclear technologies, especially how bombs affected human bodies, culture, and society.[21] But the burgeoning scholarship on war and the environment suggests that the natural world has long been an integral component of military matters, broadly defined, and therefore more work is needed that examines the intersection of nuclear technologies and the environment.[22] Recent works have begun to do that.

Mostly in the last decade, scholars have increasingly published works that more seriously consider the interplay between nuclear technologies and the natural world. These works have mostly fallen into one of two broad categories. The first category, exemplified by Jacob Hamblin's *Poison in the Well* and Mark Fiege's "Atomic Sublime," center on the social and environmental dimensions of scientific knowledge production.[23] The second category, typified by Kate Brown's *Plutopia* and Mark Merlin and Ricardo Gonzalez's "Environmental Impacts of Nuclear Testing in Remote Oceania, 1946–1996," emphasize the tremendous environmental impact of nuclear technologies as well as the attendant social costs.[24] These works and others have been quite important in helping to uncover the tripartite relationship between nuclear technologies, the environment, and human societies. But scholars have been less attentive to probing how larger issues of national policy regarding nuclear technologies might also have a significant environmental dimension. *Atomic Environments* attempts to do more than just fill that historiographical gap. Separating technology from the natural world is challenging, and rendering humans as distinct and wholly separate from either even more difficult.[25] By bringing in insights from the envirotech subfield of environmental history, which sees environment and technology as inseparable and interdependent, this book demonstrates the state-level societal connections between nuclear technologies and the environment.[26]

The rest of this book is divided into five chapters that each probe a different facet of the entanglement between environment, nuclear technologies, and policymaking. Part 1 is composed of three chapters that form a rough narrative arc about nuclear weapons. Chapter 1 situates bombs in their "natural habitat" by considering why nuclear tests occurred where they did and what testers thought those tests' interactions were with the environment, particularly meteorological phenomena. Chapter 2 reflects on nuclear fallout and argues that nuclear tests actually functioned as a massive, uncontrolled experiment in world environments and human bodies, which intermingled medicine, nuclear science, and environmental science. Chapter 3 shows how the environmental knowledge gained in the first two chapters led to nuclear test-ban treaty talks during the Eisenhower era, where the advancement of environmental knowledge and the natural world itself became crucial landscapes of conflict in a Cold War story about the creation of nuclear test detection and evasion systems.

The last two chapters make up part 2 and step away from weapons to question how other nuclear technologies and facets of the US nuclear program interacted with the natural world. Chapter 4 shows that agriculture was an essential element of the US nuclear program, not only for how well it fit modernizing trends in agricultural science, especially via its breakthroughs using radioisotopes and also the direct application of radiation to foods, but because it provided the US government a wholly peaceful aspect of nuclear energy to change the moral position of their actions. Chapter 5 shows that nuclear waste disposal was conceptualized to be just like any other waste disposal in the United States but with the added component of radiation, showing how learning about the natural world and how its systems functioned was sometimes important just to uncover the most effective ways to dump nuclear waste into it. The conclusion, finally, muses about how the mingling of nuclear and environmental science might cause us to reconsider our expectations for the inclusion of environmentalism into policymaking. Running throughout each chapter is the consistent argument that environmental science played a heavy role in the development of nuclear technologies as it bolstered executive policymaking priorities that favored institutions over environmental or human health.

Atomic Environments thus attempts to cleave through the Gordian Knot of nuclear technologies, environment, and human society. Official AEC historians Richard Hewlett and Oscar Anderson Jr. claimed in 1962, "No other development in our lifetime has been fraught with such consequences for good or evil as has atomic fission."[27] But like any technology, the determination of whether splitting the atom has been a force for good or evil certainly is in the eye of the beholder. It is not the purpose of this book to pass a moral judgment on the creation and implementation of nuclear technologies, nor on the people who did so. But, a closer look at the ways environmental science contributed to the development of nuclear technologies in the postwar United States does allow us to see what policymakers valued about scientific knowledge, research, and development. Such examination also provides a glimpse into policymakers' worldview that desired to produce nuclear technologies to protect and improve the nation but needed environmental science to do so. On one level, this book is about holding our political leaders accountable for their actions and decisions—only by understanding their mindsets and thought processes can we truly evaluate their actions. On another level, however, it is about recognizing how embedded the natural world is within our societies, even those parts of our lives that, other than their destructive tendencies, might seem to be wholly divorced from the environment. By doing so, perhaps we will be able to make better decisions about how we interact with the natural world and each other through it.

PART ONE

Nuclear Weapons Testing in
Natural and Political Systems

CHAPTER ONE

Bombs in Their Natural Habitats

"At the appointed time, there was a blinding flash lighting up the whole area brighter than the brightest daylight. A mountain range three miles from the observation point stood out in bold relief. Then came a tremendous sustained roar and a heavy pressure wave which knocked down two men outside the control center. Immediately thereafter, a huge multi-colored surging cloud boiled to an altitude of over 40,000 feet. Clouds in its patch disappeared. Soon the shifting substratosphere winds dispersed the now grey mass."

—US War Department description of the first atomic bomb detonation[1]

At 5:30 in the morning on July 16, 1945, humans first harnessed the power of the atom in the form of a bomb. The test occurred at a "remote section" of the Alamogordo Air Base in New Mexico, well away from the public. One US War Department press release described how, before the test, "Darkening heavens, pouring forth rain and lightning immediately up to the zero hour, heightened the drama." The "ominous weather" even "had a very sobering effect on the assembled experts whose work was accomplished amid lightning flashes and peals of thunder." All told, the "unusual and upsetting" weather even blocked out aerial observation of the test. This sense of foreboding is well captured in figure 1.1, with the now iconic mushroom cloud dominating the entire frame. Thus, even though there was not "assurance of favorable weather," the first atomic detonation occurred as scheduled.[2] The actual test itself presented a terrifyingly striking and yet fascinatingly horrible spectacle, as the War Department's description attested.[3] The world's first nuclear test, dubbed "Trinity," likely harmed the New Mexican desert environment just as

9.0 SEC.
N   ⊢————⊣ 100 METERS

Fig. 1.1. "Trinity," the world's first detonation of a nuclear weapon. After this test, which occurred on July 16, 1945, Robert Oppenheimer quoted from the *Bhagavad-Gita*, "I am become Death, the shatterer of worlds." Image courtesy of the United States Department of Energy.

much as the Upshot-Knothole Encore test shot did its forest as described in this book's introduction. But, owing to modern humans' fondness for forests and fears of the desert, testers paid much less concern to how desert ecology might have been damaged.

From the very outset, the interchange between nuclear science and environmental science played a considerable role in US nuclear tests. Historian Ferenc M. Szasz described the Trinity test site as an "open-air scientific laboratory" that blurred the lines between nuclear science and the natural world.[4] Figure 1.2 helps capture that sentiment, catching the Trinity detonation at an odd time, only 25 milliseconds after detonation. It took precise camerawork to capture the blast before it could form the infamous mushroom shape. The official War Department public release described the blast in relation to the natural world, particularly mentioning the weather and mountains as a natural setting. In fact, weather had been a significant factor in the tests because summer thunderstorms could come upon the site so quickly and unexpectedly. Jack M. Hubbard, the chief meteorologist for the project, described how

Fig. 1.2. "Trinity" 25 milliseconds after detonation. At this point, the fireball is about 600 feet wide. Image courtesy of the United States Department of Energy.

the operation date was "set from above," and after that it became necessary to tailor the experiment to the expected wind and weather conditions. Moreover, questions existed about whether a blast would produce rain or affect the winds.[5] Future nuclear tests also depended upon environmental science as the United States continued to test nuclear weapons to maintain and increase its nuclear superiority over the rest of the world, particularly the Soviet Union. The Soviets tested their first nuclear bomb in 1949, and after that US testing became even more imperative from a military perspective.

This chapter outlines the Atomic Energy Commission's (AEC) development of the US nuclear arsenal by taking the sort of banal environmental elements historians typically overlook and reinterpreting otherwise familiar tests through them. Trying to write a brief comprehensive history of US nuclear testing would be completely impossible, and this chapter does not attempt to do so. What it does do is roughly chronicle US tests and pull out some of the moments that most captured the attentions of executive policymakers. Scholars have well studied how weapons detonations had tremendous effects on both local and global environments and that the study of the natural world benefited tremendously from military patronage. But this chapter demonstrates the degree to which environmental science became ingrained in US nuclear testing and helped shape the scientific and political understandings produced by testing.

The ways in which environmental science meshed into nuclear tests were multitude. Geology, biology, geography, meteorology and other disciplines helped uncover the different ways bombs affected the biosphere, lithosphere, hydrosphere, and atmosphere. Indeed, executive policymakers drew on these disciplines to locate and craft functionally "natural" habitats for the testing of atomic bombs. In exploring the militarized landscapes that marked the Cold War, historians Simo Laakkonen, Viktor Pál, and Richard Tucker contended, "societies and militaries not only destroy landscapes, they also produce them."[6] That idea is especially evident when considering nuclear weapons testing that, in addition to having obvious negative effects on the natural world, also depended on very particular environmental locations and conditions to produce the most accurate and useful scientific data. To those planning nuclear tests, certain environments proved much more conducive to successful nuclear tests, meaning decision makers conceived of the bombs belonging in some places more than others. Uncovering how policymakers navigated the connections between bombs and their corresponding natural environments over time thus links together the seemingly disparate episodes explored in this chapter.

Historical scholarship on the interaction between nuclear weapons and the environment has principally focused on how tests damaged environmental health.[7] Even though this chapter seeks to push beyond such understandings, those works are not wrong to do so. Nuclear bombs have had incredibly deleterious effects on both humans and the natural world.[8] Moreover, critics of atomic weapons testing, of whom biologist and ecologist Barry Commoner is perhaps the best known, did important work in showing the world exactly how dangerous nuclear weapons testing could be. Commoner wrote in 1958, "In part, our present troubles derive from the unequal pace of the development of physics and biology." By this, he meant that, while nuclear science had advanced enough "to explode great quantities of radioactive materials into the atmosphere," biology was limited in how it could help humans deal with the attendant radioactive fallout.[9] Commoner led the charge for scientists to become more involved in politics and use their knowledge to help end nuclear testing. His work helped make the world a safer, healthier place for all living beings on the planet. As historian Michael Egan argued, in many ways the modern environmental movement was a response to Commoner's work because he showed society that a clean environment was more than "just a desirable commodity; it was a social necessity."[10] And yet, moving beyond such understandings is paramount to creating a more nuanced understanding of how and why testing occurred in the first place. Clearly tests produced innumerable deleterious effects on human bodies and the world around them, and therefore a reckoning of the decision process that led to those tests is necessary. Doing so requires starting with the earliest nuclear tests.

The Manhattan Project reminds us a quirk exists in the history of nuclear weapons—although the most well-known nuclear blasts occurred during World War II, the overwhelming bulk of atomic weapon detonations occurred as tests during times of relative peace. By the time World War II ended, the United States had detonated the world's second and third nuclear bombs.[11] President Truman learned of the first wartime use of an atomic bomb from a telegram, which informed him that reports on the "Big bomb dropped on Hiroshima . . .indicate complete success which was even more conspicuous than earlier test."[12] Such language implied that the bombing of Hiroshima was a test just like the world's first nuclear bomb detonation at Trinity.[13] Whether bombs dropped as acts of war truly can be considered tests or not is beyond the purview of this book, but the mere comparison implicit in the referenced telegram demonstrates the entangled nature between testing and scientific knowledge.

The physical origins of nuclear bombs—and all nuclear technologies for that matter—quite literally reside in the earth. Nuclear technologies began with the procurement of radioactive ores through mining, deemed important enough that the AEC recruited a Director of Raw Materials in 1947.[14] The usable component of any mined radioactive ores represented only a miniscule fraction of the total extracted minerals, even in uranium-rich locations (a 1949 estimate placed the ratio of uranium to other materials in the Earth's crust at only roughly 6:1,000,000).[15] Processing facilities turned raw ore into usable fuel for implementation as a variety of nuclear technologies, especially used in nuclear weapons and nuclear power production. At that point, atoms from the processed ores could be fractured to release tremendous bursts of energy, meaning that humans worked with the ores on multiple scales both conceptually and physically. But even from the outset, securing radioactive ores formed a crucial component of the US nuclear program.

A mere day after the United States devastated Nagasaki with the world's third atomic bomb, President Harry Truman received a letter that implored him to make securing uranium national policy. In writing that letter, Abe Fortas, the Acting Secretary of the Interior, demonstrated that even early on policymakers recognized the importance of controlling the natural world to developing nuclear technologies. Without a steady supply of uranium, the United States could never continue its nuclear program. Fortas contended, "The recent use of radio-active mineral substances as an agency of destruction for war use and the strong possibility that they may be adapted to new and revolutionary uses in time of peace will undoubtedly result in an intense search for and acquisition of lands containing such substances." Because of the inherent importance of any lands containing uranium, the acting secretary believed such lands should not be permitted to fall into private hands in any way. Fortas then proposed that the president issue an executive order withdrawing all public lands containing radioactive source materials from any sort of purchase availability to the public.[16]

A month later, President Truman signed executive order 9613, which removed all public lands and future lands of the United States that "contain deposits of radio-active mineral substances, and all deposits of such substances" from disposal under all public-land and mining laws.[17] Executive order 9613 highlighted two aspects of the US nuclear program: the system was highly dependent on radioactive ores, directly tying the fate of nuclear research to the land and the materials that could be mined from it, and the development of a nuclear program of sweeping extent began to develop fairly early in the country's nuclear program, even before the end of World War II.[18]

The United States' first major cache of radioactive ores—1,250 tons of high-grade uranium from the Belgian Congo—was shipped to a Staten Island warehouse by Belgian industrialist Edgar Sengier. The Manhattan Project promptly bought all of that uranium and then started trying to acquire even more, mostly staying within the nation's borders at first.[19] But the trend of the US nuclear program expanding its geographical reach—the nation's search for radioactive ores reached across most of the globe, particularly into Africa—only increased over time.[20] After World War II, US nuclear tests occurred primarily at two locations—in the US Southwest, like the world's first test, and at Pacific Ocean locations near the Marshall Islands (about two thousand miles southwest of Hawaii). The United States' first postwar tests occurred in July 1946 in the Pacific. But those tests caused issues in the United States before they were even conducted.

Early concerns about the United States' first postwar tests were particularly manifested in fears of environmental damage. In November 1945, Congressman Schuyler Otis Bland (D-VA), chair of the House Committee on Merchant Marine and Fisheries, sent a letter to President Truman expressing concern about possible environmental damage from planned tests both underwater and in the air above the Pacific (these tests were eventually conducted in the summer of 1946 as Operation Crossroads). Bland worried that the experiments could carry "serious implications with regard to commercial fisheries" and destroy not only incredible amounts of fish but perhaps whole fisheries. He continued, "Because of oceanic circulation the effects of any induced radioactivity might easily cause great harm to the aquatic resources over wide areas." Therefore, even though Bland was not suggesting that the tests be stopped, he thought the War, State, and Navy Departments should contact the Department of the Interior to ensure that the selected testing locations would endure as little damage as possible. He argued that Interior representatives should participate and view the effects of tests on marine life. Bland ended by saying that he could "see that an immense damage may be done if proper protection is not afforded the fisheries, and I sincerely trust that everything possible may be done to protect the fisheries of the world."[21] Truman responded the next day with thanks for Bland's letter, explaining that such tests were "only in the conversation stage so far." The president also assured Bland that "every precaution will be taken if the experiment does take place to prevent any undue injuries to fish."[22]

Setting aside the thought that the fish might have been due some amount of injuries, the beginning of a pattern had emerged even before the United States conducted its first postwar nuclear weapons test. It is unclear the exact level of concern that Bland and Truman actually held for the wellbeing of fisheries. Instead of focusing on ecological health, both men thought the most promising approach to the issue was to protect commercial resources and economic interests. Appropriate bomb environments were thus understood as those that would not endanger US commercial interests.

In July 1946, in spite of protest from Bland (and many others) directed at the president, the United States conducted its first postwar nuclear tests, detonating two bombs in the Pacific Ocean on the Bikini Atoll, a small lagoon in the Marshall Island chain that became a de facto colonial holding for the United States post–World War II.[23] Testers intended for this series, Operation Crossroads, to help show what a nuclear blast would do in an aquatic environment to ships and their crews, and therefore had ships moored in the atoll's lagoon, spread out at predetermined distances.[24] In addition to allowing the United States to conduct oceanic tests, Bikini also had the advantage of being several thousand miles away from the US mainland. The first test of this Crossroads series was codenamed "Able."

Fig. 1.3. Operation Crossroads "Able" shot, detonated July 1, 1946. Able was the first post–World War II nuclear test. In the second postwar nuclear weapon test, the underwater detonation spewed copious radioactive water into the air and carried it out in hundred-foot-tall waves. Image courtesy of the United States Department of Energy.

By all accounts, detonation of Able (figs. 1.3–1.4) went mostly as planned, other than going off 1,500–2,000 feet west of the assigned target. Its intensity "approached the best of the three previous atomic bombs," and it detonated at the planned altitude, a few hundred feet above the sea. Many of the ships surrounding the blast showed considerable damage. Exposed personnel on those ships (had there been any) would have experienced high casualties, but those sheltered "would not have been immediately incapacitated by burns alone" no matter what happened later from the lethal doses of radiation they would have received. "In general," the report on test Able said, "no significant unexpected phenomenon occurred, although the test was designed to cope with considerable variation from predictions. There was no large water wave formed. The radioactive residue dissipated in the manner expected. No damage occurred on Bikini Island, about three miles from the explosion center." The test went well enough that evaluators claimed, "The importance of large-scale research has been dramatically demonstrated."[25]

The second Crossroads test, codenamed "Baker," did have unexpected environmental phenomena. Baker (figs. 1.5–1.7) was an underwater shot that

Fig. 1.4. Closer aerial view of Operation Crossroads "Able" shot. The lagoon setting of Operation Crossroads is especially evident in this photograph. Image courtesy of the United States Department of Energy.

Fig. 1.5. A view of Operation Crossroads "Baker" shot. The juxtaposition between beautiful, natural features in the foreground and the iconic mushroom cloud in the background reminds us that nuclear tests were always situated within an environmental context. Image courtesy of the United States Department of Energy.

Fig. 1.6. Operation Crossroads "Baker" shot detonated July 25, 1946. From this angle it is easy to see the ships moored at varying distances around the blast epicenter and the considerable spray of water. Image courtesy of the United States Department of Energy.

caused unexpectedly gigantic waves, which, even 1,000 feet from the center of the detonation, were 80–100 feet in height. The bigger problem, however, was the incredible radiation from the test and the unexpected form it took. The preliminary report on the subject to President Truman said, "Great quantities of radioactive water descended upon the ships from the column or were thrown over them by waves." Such "highly lethal radioactive water" made it unsafe for any inspection teams to board the moored ships for days. Beyond the ships physically destroyed by the blast (testers used a similar setup to the Able shot, especially well seen in fig. 1.6), the surviving ships were effectively scuttled while still afloat. Telling the president that it was "impossible to evaluate an atomic burst in terms of conventional explosives," the report generally described the Baker test as incredibly destructive. It summed up the radiation contamination of the ships by saying they "became radioactive stoves, and would have burned all living things aboard them with invisible and painless but deadly radiation."[26]

These two tests show an evolving understanding about the relationship between testing atomic bombs and scientific knowledge about the environment.

Atomic bomb tests were, at their core, scientific experiments. Hence while the first test, Able, likely spewed radioactive materials into the air, airborne contamination went mostly unmentioned in reports because the experiment was not designed to measure for such. The radioactive water from the Baker shot, on the other hand, occupied a central place in analyses because its effects were readily detected within the existing experimental framework attempting to test for radioactive contamination of the moored ships.

The different reactions to Able and Baker show the importance of environmental science to test evaluations. While evaluators might have been able to avoid talking about a radioactive column of air or minimize its effects in their reports because they did not detect it, a tidal wave of radioactive water proved too much to ignore, as the Committee to Observe the Atomic Bomb Tests estimated the radioactive watery spray covered 90 percent of the target array (moored ships and submarines surrounding the blast). The committee asserted this was the equivalent to exposing the area to "many hundred tons of radium." In addition, meteorology played an important role in judging experimental results. When evaluators deemed weather conditions for the Baker shot were "perfect," they demonstrated that nuclear tests changed the criteria by which testers evaluated the weather. The weather, in this instance, had to be understood through the aims and desires for nuclear tests. In addition, had weather conditions not been "perfect," any test results would have been mediated through that evaluation of less than perfect weather. After Operation Crossroads wrapped up, the committee looked at the possible effects of nuclear bombs, especially including the nuclear tidal wave the Baker shot produced, and concluded, "The Bikini tests strongly indicate that future wars employing atomic bombs may well destroy nations and change present standards of civilization."[27] Weather, water, and air were not merely the settings for nuclear weapons testing, but instead became the experimental mediums through which humans understood and evaluated the tests' successes, failures, and possible impact on civilization and nations.

Of course, even though the importance of environmental knowledge to these tests is clear in hindsight, it does not mean that evaluators thought of either the tests or their reports in explicitly environmental terms. Instead, the assessments judged the incredible waves of radioactive water to be an "impressive result" of the Baker test and not necessarily an environmental phenomenon—the wave was more a scientific product than part of the natural world. The final report of the Joint Chiefs of Staff declared that the tests at Bikini "provided data essential to future military planning, giving bases for the calculation of the conditions under which the maximum destructive effects of an atomic explosion will be obtained against various types of land and water targets and against living organisms." Testers had used live animals as test subjects on the moored ships during the tests, but even if effects on animal biology comprised an important aspect of the tests, evaluators thought of the affected beings purely in military terms. Animals on those ships were important living creatures to testers because they represented hypothetical future bomb targets, not because the animals' lives had any inherent value or worth. Just as bombing pilots might talk about "the hit," but not necessarily the destructive aftermath of their bombs, test evaluators proved more concerned with the effects their bombs might have on targets than they necessarily were with the natural world and creatures in it.[28] Environmental science was more important for how it could advance nuclear science than for understanding the health of studied environments, as demonstrated by the more than thirty thousand pages of "detailed technical reports" containing information on the "military aspects of radiological safety, and those relating to oceanography, meteorology, and marine and island biology and geology" and not anything related to protecting ecological health.[29] To evaluators, nuclear testing sites were not spaces where nuclear technology interacted with the environment. Instead, as historian Ferenc Szasz suggested, they were open-air testing laboratories where the natural world became an integral part of the testing process and not merely the setting for destruction.

In addition to textual evidence, photographs of the events help demonstrate the key role the natural world played in Operation Crossroads. Figure 1.4 provides tangible evidence that these tests did not occur "in the middle of nowhere" but instead nestled right in the middle of the Bikini Atoll. Two scientists from what is today the Smithsonian Museum of Natural History were among a team that conducted biological surveys of the islands, noting all the animals killed by the blasts (especially fish after the underwater Baker test). Additionally, the US government forced 167 native Bikini residents to relocate, never to return to their homes due to increased radiation.[30] The islands and natural world mattered to everyone involved, and figures 1.3–1.7 thus demonstrate a real sense of place that is not easily gleaned from documentary evidence. Figures 1.5 and 1.7 are clearly taken from one of the Bikini islands, showing a mixture of palm trees and clean sandy beaches on one hand, and huts, docks, and boats on the other; a juxtaposition of human creations onto what was conceived as a natural island. The atoll itself therefore proved crucial to setting up cameras and other instruments in these tests. It

Fig. 1.7. Operation Crossroads "Baker" shot, again. In this photograph, both the moored ships and nuclear test (at the time the height of technology) combine with the palm trees (reminiscent of a tropical paradise) to create an almost paradoxical scene. Image courtesy of the United States Department of Energy.

is not going too far to say that these tests were despoiling a tropical paradise, but in some ways that was exactly the point of Operation Crossroads. Testers did not know exactly what would happen if they detonated nuclear bombs under these circumstances, and therefore the island environments helped answer their questions. Destruction of tropical splendor may have seemed like an unfortunate side effect of Operation Crossroads, but it was, as a matter of fact, a partial goal.

Operation Crossroads also demonstrated that Alamogordo, New Mexico, was not a suitable site for all US nuclear weapons tests, but Bikini was not always a suitable habitat either. Therefore the AEC's late-1940s establishment of "a proving ground in the Pacific for routine experiments and tests of atomic weapons" not only showed the importance of testing to the commission but also demonstrated how policymakers increasingly involved environmental knowledge in their decisions about nuclear tests. The commission declared that tests "must be a routine part of any weapons program," and the president later approved that statement for inclusion in an AEC report.[31] The federal government would later search for a continental site for the top secret Project Nutmeg, even considering the Outer Banks of North Carolina as potential testing grounds before settling on the site in Nevada.[32]

The required characteristics for test locations melded geopolitical and environmental considerations. One summary listed those requirements: (1) protected anchorage six miles in diameter; (2) at least three hundred miles from urban areas; (3) less than one thousand miles from a B-29 base; (4) a region without "violent storms"; (5) predictable ocean currents with high dispersion rates, and fast surface currents that avoid "fishing areas, steamer lines and inhabited shores"; (6) a certain minimum distance from the continental United States; (7) an unpopulated area; (8) an area owned/controlled by United States; (9) a temperate tropical climate. In the end, planners selected Bikini as the best area to fit into the qualifications for early tests. The resident "population [was] less than 200 and can be readily moved to another atoll" (whether they wanted to or not). And, in deference to the Department of Interior's wishes, the AEC studied the location with respect to the effects tests there would have on fisheries.[33] Long before any tests ever occurred, without any real concern for environmental health or protection of the natural environment, in choosing their ideal testing site policymakers explicitly considered the natural world and how bomb testing would affect the environment and vice versa.

For larger tests, AEC higher-ups used environmental and scientific reasoning and eschewed Bikini Atoll in favor of establishing a permanent Pacific Proving Grounds at Eniwetok Atoll. The two locations are roughly one hundred miles apart, both in the Marshall Islands chain. The AEC needed a "suitable" area to test bombs, and Eniwetok gained the nod because Bikini did not have the sufficient land surface to accommodate the instrumentation for proper scientific observations. Also, only 145 residents called Eniwetok home, and "very important from a radiological standpoint, it is isolated and there are hundreds of miles of open seas in the direction in which winds might carry radioactive particles." Perhaps more important than keeping human bodies safe was whether the meteorological conditions and winds "might make it much easier for a foreign observer to obtain samples of a radioactive cloud and possibly to inform information concerning the composition or efficiency of an atom bomb."[34] Eniwetok's natural features therefore proved crucial for its selection. Not only did the Atoll provide enough land area but it also held the geographic advantage of being far away from other peoples (its residents and their connection to their homeland were out of luck, though, in this health and safety calculus). And local meteorology could help provide a national security boost by keeping bomb composition secret.

A press release on Eniwetok's selection even declared, "All test operations will be under laboratory control conditions," meaning a few decisions and judgments on the Atoll's natural features, along with instrumentation, could turn Eniwetok, in the minds of US policymakers, into a laboratory.[35] As historian E. Jerry Jessee explained, Bikini's "relative disconnectedness from pelagic processes rendered it as a relatively self-contained space, not unlike a modern laboratory," leading Rear Admiral William Parsons, the deputy commander for the tests, to call Operation Crossroads the "largest laboratory experiment in history."[36] It is thus no surprise that historian of science Robert Kohler has argued that there is "not a simple line" that separates laboratories from the field but instead there exists "a cultural zone with its own complex topography of practices and distinctions." Such a statement makes sense in this context, as does his later comment that experiments are "the better, or even the only, way of knowing nature."[37] That attitude pervaded US nuclear testing.

Other nuclear weapons tests showed a deepening connection between the natural world, testing, and environmental science. In 1948, the United States conducted a series of tests named Operation Sandstone at the Pacific Proving Grounds. Sandstone consisted of three nuclear test shots, and after each one, testers sent in tanks, remote-controlled from helicopters, to take soil samples for analysis. However, each time at least one tank bogged down in the soft soil and another one had to be sent in to do the first tank's job. After nearly a week, soil radioactivity declined enough that testers felt confident they could send in teams to recover stalled tanks without exposing them to excessive radiation.[38] Testers proved quite aware of the on-site radiation present after each nuclear blast, even if their primary concern certainly was about human health and not the condition of the environment. But even with concern for radiation, something as simple as soil density could throw off the plans of even the best-prepared evaluators.

In the fall of 1950, US policymakers began discussions about conducting underground tests on Amchitka Island, located in the Aleutian chain, largely because it provided a different potential habitat for US nuclear weapons that would create new and different scientific understandings.[39] Though tests at the site ultimately would not happen until well after the end of Truman's presidency, discussions reached a critical enough stage that Truman approved testing during fall 1951 at Amchitka.[40] What makes the case of testing at Amchitka worth studying is that scientific findings about the environment eventually scuttled the test series even if concern for environmental health did not factor into the decision.

The planned tests at Amchitka further demonstrate the connections between environmental science and nuclear weapons testing as the two created a need for each other, enmeshing the natural world into the development of the US nuclear arsenal. Evaluations of the eight atomic bomb tests to that date had provided the United States with a great deal of information on the effects of bombs detonated at heights up to two thousand feet in the air and, in the case of the Baker shot during Operation Crossroads, even underwater. Yet no information existed on the effects of a nuclear bomb detonated underground. If the United States needed a penetrating atomic weapon to attack "particularly well constructed or deep underground targets," the data at hand would not be sufficient to construct such a device.[41] US decision makers consequently sensed a need for tests that would help them better understand the relationship between nuclear weapons detonations and the lithosphere.

Site selection held particular importance in discussions of underground testing, and choosing a site for the test hinged on nine factors that combined issues of environment, politics, and logistics: safety, sovereignty, security, public relations, climate, geology, cost, accessibility, and size. While testers did not think that underground nuclear testing would be particularly dangerous (at least in comparison to other atomic tests), they still did not want to hold the tests in the United States, believing "Certain remote areas in Canada and other areas within the Commonwealth, such as Australia, offered some advantages." These remote Canadian sites made the cut over many other sites; the Caribbean had too many people, the Pacific Proving Grounds at Eniwetok did not have the correct size or geology, and evaluators discarded many Alaskan sites "because of inaccessibility, extreme climate, unsatisfactory geology and the considerable number of trappers and prospectors."[42]

A combination of human and environmental factors consequently intermingled to create the determining criteria for the site selection of this test. The selection of the eventual site on Amchitka showed this well, as "Careful consideration of the several isolated, uninhabited islands toward the outer end of the Aleutian Chain led to the determination that Amchitka Island is the only site that satisfies all of the established criteria to an acceptable degree." Even though Amchitka was completely uninhabited (because the US government had removed native inhabitants during World War II, quite similar to the colonialist relationship the United States took to the Marshallese), the island had the infrastructure necessary for testing, leftover from World War II.

Some factors grounded in the natural world worked against the site, though, such as the island's mostly "bad" climate and the Department of the Interior's strong desire to preserve indigenous wildlife. A selection memo further explained, "Rain and fog predominate in the summer and snow and high winds in the winter." Such concerns could be worked around, however, as "for a short period in May and in a longer period in September and October, the weather can be expected to be moderate." Hence even with a few environmental problems and security concerns because the island was so close to the Soviet Union, policymakers deemed Amchitka Island "the only site presently available that reasonably satisfies all the criteria established for the safe conduct of an underground and surface atomic test." A report also noted that prevailing winds in the region from west to east meant that the USSR would not be able to detect the tests by radiological means. In this case, the site had prevailing winds working in its favor because national security was deemed more important than reducing the radiation exposure of US and allied citizens.[43] Even with some legitimate concerns, Amchitka, due to its geography and environment, represented the most ideal site for the United States' proposed underground tests.

Site selection at Amchitka was, however, challenged by significant criticism emerging from the Department of Interior site due to concerns about wildlife protection. Dale E. Doty, assistant secretary of the Department of the Interior, wrote that he and his department protested using the island for testing because it represented "the principal concentration center for the total existing population of sea otter which had been brought to near extinction during Russian occupation of Alaska and which is now being restored and re-colonized over a part of its former range under the close protection by the Fish and Wildlife Service." US officials used the Amchitka herd for stock as part of the transplanting and management program, and though there were otters elsewhere, the herd had only increased on Amchitka, making the value of the site clear. This is not to say that Doty believed in purely preservationist ideals— he also argued that the value of each otter pelt would average around $1,000 each, with some topping $2,500. Keeping the site viable for otters would produce "revenue to the Government, once the resource is restored to a production basis, [of] hundreds of thousands of dollars annually."[44] Additionally, the fact that Soviets had nearly extirpated the otters meant that the United States could claim the moral high ground in the early Cold War.

Doty and the Department of the Interior thus objected to testing on Amchitka for several reasons, all related to the otter population. First, testing personnel entering the site "would undoubtedly provide opportunity for the molesting and killing of animals." And since the beaches of the island were situated within the danger area of the test, the blast itself, along with "falling debris, flash, and possibly direct radiation," caused a good chance of harming the animals. The potential for long-lived radiation also worried the Department of Interior. Because of this, from the standpoint of the Department of Interior's responsibilities toward the sea otter and waterfowl populations, "it would hardly have been possible to have chosen a more objectionable area than Amchitka." Doty argued that certain provisions needed to be incorporated into test programs if the operations ever proceeded. All laws needed to be followed to ensure "the maximum possible protection of the sea otter from poaching or from any unnecessary disturbance or molestation," and the Fish and Wildlife Service needed money to trap and transfer many otters over the winter to safer areas. The latter would be costly and require "considerable logistic support" from the US military. Even considering all these precautions, Doty still believed that the tests should happen elsewhere, as Amchitka was the only place otters had recovered as well as they had.[45]

In many ways, the dispute between the Department of Interior and planners for tests on Amchitka Island demonstrate the conflicting purviews of different bureaucratic actors. While the Interior Department felt obligated to protect US wildlife, policymakers in the White House wanted nuclear tests to improve the nation's nuclear arsenal. Differing agency responsibilities led to different notions about the best way to protect the United States. In the end, testers noted all objections but decided to go on with the test and merely to work with the Department of the Interior "to preserve the indigenous wild life inhabiting the island."[46]

While managed otter populations alone could not deter those wanting to test on Amchitka, the area's natural formations could. Before the proposed fall 1951 testing date, reports surfaced that "detailed exploration of [Amchitka Island] revealed geological conditions less favorable than preliminary surveys had indicated." Analysis of the newly discovered "geological conditions" indicated that any data gained from tests on Amchitka would be less accurate than initially believed. This caused policymakers to rethink their plans and instead believe that perhaps a continental site might be better, as more "favorable geological and meteorological conditions are known to exist at several possible continental test sites than at Amchitka Island." Ultimately, even though it was later used in a different test series, Amchitka was jettisoned for the Nevada Test Site with the recognition that a continental site would also clear up many logistical problems.[47]

The case of Amchitka therefore helps demonstrate the selective inclusion of environmental science into planning and how some environmental factors simply mattered more to nuclear policymakers than others. While factors like the survival of sea otters or what might be perceived today as environmental health might have mattered to the Department of the Interior, to policymakers testing nuclear weapons such considerations were deemed less important, at least in terms of what could be worked around and what could not. What proved crucial, on the other hand, were aspects, both geological and broadly environmental, that might have influenced the accuracy of data gained during US nuclear tests. The process of selecting Amchitka as an experimental test site therefore shows how decisive certain (not all) environmental concerns could be to US nuclear policymakers.

After the Amchitka discussion, the AEC and other executive policymakers settled into a pattern where the most common ways they interacted with the natural world surrounding nuclear tests centered on radiation from blasts and interactions with the weather, reminding us that nuclear bombs' habitats included the atmosphere. Radiation damage on the natural world was, for the most part, downplayed by those in power. In 1951, the AEC noted that high air burst weapons tests showed that residual radiation was not a problem at ground level, but on blasts close to the ground significant radiation contamination existed. Of course, since such detonations destroyed everything for at least a 300–400 yard radius, residual radiation was something of a moot point.[48] And a March 1952 memo from the AEC to the White House claimed that upcoming tests might produce "some off-site radiation above normal levels, but far below levels harmful in any way to humans, animals or crops."[49] Yet one of the more interesting interplays between atomic weapons and environmental science involved decisions and considerations about meteorological phenomena.

After the Second World War, meteorology increasingly became a field developed by military patronage, as weather prediction mattered a great deal to the US military. Before World War II, civilians controlled meteorological research during peacetime with the military typically taking over during times of war. World War II changed that relationship, however, with the military maintaining control of meteorological research funding after the war. Because of that, while postwar academics pursued theoretical knowledge, the

US Weather Bureau concentrated on improving forecasts. Developed in large part from military funding, numerical weather prediction became, in the estimation of historian Kristine Harper, the twentieth century's "most important scientific advance in meteorology."[50]

In the early 1950s, both the AEC and White House paid significant attention to how weather might influence atomic bomb tests and how those tests possibly affected the weather. Before and after any nuclear blast, testers made significant meteorological measurements to ensure that proper conditions existed for those tests. After one test, an Air Force "group of 2,400 made weather observations, and operated experimental aircraft including radar-directed 'drones' to collect observations in and near the radioactive clouds that follow atomic explosions."[51] In the aforementioned March 1952 memo, the AEC representative said that precautions for preventing excessive radiation included "cloud tracking and sampling" by the Air Force and cooperation with around one hundred Weather Service stations.[52] These two pieces of data started to show the importance of weather to those testing atomic weapons, but do not fully show the true weight of meteorological effects on the process.

The January 1953 semiannual AEC report to Congress contained a great deal of focus on weather and atomic bomb tests. To begin, the report explained the precautions taken during weapons tests to prevent "hazard to the public from blast or fall-out." To do this, the AEC constructed a national system to monitor "fall-out radioactivity." One of the best ways both to limit the spread of fallout and to gain accurate test results was to make sure that proper weather conditions existed before tests. Before each test, predetonation forecasts began seventy-two hours in advance. If those predictions were still favorable twenty-four hours before a scheduled test, the operational sequence began. If weather proved unfavorable, the test might be canceled. Since wind and rain were known to affect fallout distribution dramatically, the weather formed a crucial part of ensuring tests were as safe as possible. The report also claimed that the "intensity of blast waves at any locality depends more upon various weather phenomena than upon the energy yield of the detonation."[53] Weather was not just a secondary concern when planning and conducting tests—it mattered a great deal to planners who wanted the tests to be as safe as possible and the data gathered from tests to be accurate.

For a time, significant concerns also existed about whether atomic detonations themselves might affect the weather.[54] At a May 1953 meeting of the AEC commissioners, Gordon Dean, chair of the AEC at the time, questioned whether adverse weather conditions following detonations at Eniwetok could be attributed to those tests. A report had claimed the bad weather "appeared to have been caused by the shots." Photographs showed that heavy clouds and squalls developed after the test shots, along with a series of high-altitude storms (around forty thousand feet high). The committee eventually decided "weather conditions prior to the shot time were favorable to rain, and the large vertical disturbances caused by the blast seemed to have 'triggered' the storms which began at Eniwetok and spread north and west over an area of 250,000 square miles." But they also commented that "meteorological experts" had not discerned "any relationship between the recent weather conditions throughout the U.S. and the Nevada tests."[55] Even though experts had not decided on a link between tests and weather, many in the public had.

Several letters from civilians can show how many citizens connected atomic bomb tests and meteorological phenomena. One letter to Sherman Adams, Eisenhower's White House Chief of Staff, talked about the author's "very dear friend," a seventy-four-year-old farmer. That farmer's crops were six weeks late, and he believed, "as many others do, that the atomic bomb [was] responsible for it." The farmer wanted the blasts "postponed for a while so the nation wouldn't starve to death."[56] Another man, in a letter to James C. Haggerty, Eisenhower's press secretary, claimed that tests needed to stop. This letter claimed, "Due to atmospheric changes due to the high explosives of the atomic weapons," there now existed problems "with the atmospheric conditions in our country resulting in tornadoes where they have either never previously occurred, or—where tornadoes have previously been experienced—now being of unusual intensity."[57] While neither environmental science at the time or at present would support such assertions, the letters do demonstrate examples of public pressure exerted on White House policymakers over worries about a connection between the natural world and atomic bombs.

In June 1953, the AEC commissioners again met to discuss the effects of nuclear tests on weather, particularly because of "numerous charges" in the press saying that tests at Nevada Proving Grounds had caused "unusual weather conditions in parts of the United States."[58] During previous tests at Eniwetok, weather conditions after tests had been in accord with pretest meteorological predictions. With the evidence presented, the commissioners reckoned that the disturbances after several tests, which included "rain squalls over the ocean [and] small storms, but no winds of hurricane force," might have been caused by blasts. A military representative, along with a UCLA-based scientist, claimed that there had been similar air circulations following tests in Nevada to those at the Pacific Proving Grounds, but the continental tests only caused disturbances for a few minutes and lacked sufficient moisture in the desert to create storm conditions. They continued, "No material in the bomb debris could cause rain or a tornado. It was possible for a tornado to be 'triggered' by external conditions, but it needed moisture as a fuel to become selfsustaining." Instead, the unusual number of tornadoes that spring "could be attributed to an unusual pressure condition forcing moist Gulf air across the U. S. at high level until it came in contact with a cold air mass coming down from Canada, and that by no mechanism known was it possible for the tornadoes to have been caused by the Nevada tests." In the end, the commissioners decided that they needed to respond to public views, especially the charges that tests at the Nevada Proving Grounds had caused tornadoes.[59] Scientific understandings about the relationship between bombs and weather may have assured the AEC commissioners that nuclear weapons could not cause violent storms, but these did not convince everyone.

No matter the official position, public perceptions of atomic bomb tests causing severe weather proved so strong that Representative Edith Nourse Rogers (R-MA) introduced a series of Congressional resolutions requesting that the AEC provide information about the connection between nuclear tests and the weather. Senator William Langer (R-ND), in a related move, offered a resolution that proposed that no further atomic tests could be held in the continental United States.[60] The issue continued to be prominent for several years, as the AEC held a conference in 1956 on "Possible Effects of Nuclear Explosions on Weather."[61]

Whether atomic tests affected the weather or not, nuclear weapons tests perhaps became more important when Dwight Eisenhower assumed the presidency on January 20, 1953. As a staunch fiscal conservative, Eisenhower's term showed him deeply committed to the responsible use of the nation's financial resources. Of course, as the former leader of Allied forces during World War II, he also focused heavily on US military commitments and issues of national security. To balance both financial and military considerations, Eisenhower devised a foreign policy plan called the "New Look." This strategy sought to use strategic nuclear weapons in lieu of conventional military forces to deter the USSR and Soviet bloc countries from attacking the United States and its allies. In short, Eisenhower figured it to be cheaper to create a nuclear stockpile than to train, equip, feed, and supply a large standing Army. Nuclear testing formed a crucial and necessary part of this "New Look" policy as improving and increasing the US nuclear stockpile meant, under this logic, keeping the nation safe in the most cost efficient way.[62]

Exemplifying this general faith in nuclear technologies, AEC Commissioner Thomas E. Murray proclaimed in a 1953 commencement address, "To my mind (to paraphrase Churchill), never was so much owed by so many to such a small amount of material [uranium]—deployed in the defense of freedom—material which the world was unaware of, so short a time ago, as when you graduates were in grammar school."[63] With this strategy in mind, under Eisenhower the United States continued previous research into atomic weapons and developed bombs of previously unfathomable power.

The hydrogen bomb, a weapon that incorporated the fusion of atoms (not principally fission like previous atomic bombs), launched the world into the thermonuclear age and at the same time radically altered the scale of potential environmental change from atomic weapons.[64] Lewis Strauss, chair of the AEC, elucidated on the thermonuclear tests at President Eisenhower's March 31, 1954, official press conference. Strauss had recently visited the Pacific Proving Grounds to view the second part of a series of tests of thermonuclear weapons. He explained that after the Soviets had detonated their first atomic bomb in August 1949, US military leaders had decided that the United States could only maintain its nuclear superiority over the Soviet Union with either a significant quantitative edge in bombs or by developing something greater than existing fission weapons "by a degree of magnitude comparable to the difference between fission bombs and conventional bombs." Therefore, in 1950, President Truman had asked the AEC to start making a hydrogen or fusion bomb—a thermonuclear bomb. The United States tested a prototype at Eniwetok in November 1952 and the Soviets tested a similar weapon in August 1953. In March 1954, however, as part of Operation Castle, the United States tested what was easily the biggest nuclear device the world had seen to that time.[65]

That early March test shot, code named "Bravo," ended up being several times more destructive than expected and produced significantly more fallout than anticipated. Within a second, the blast had created a fireball nearly three miles in diameter and dug a crater more than a mile wide and two hundred feet deep into the Bikini Island reef. Observers reported seeing the test at least 250 miles away, and it rattled windows on Rongerik Atoll about 155 miles away.[66]

Strauss, in contrast, stressed that the test did not get out of control. Even when badgered by the media about whether other tests might, he responded, "I am informed by the scientists that [a test getting out of control] is impossible."[67] Further emphasizing that the test had not been out of control as suggested, Strauss argued that the AEC "has conducted the tests of its larger weapons away from the mainland so that the fall-out would occur in the ocean where it would be quickly dissipated both by dilution and by the rapid decay of most of the radioactivity which is of short duration." As explained previously, this is why the United States conducted previous tests at Bikini in the Marshall Islands—it has good winds from February to April that would supposedly blow any fallout away from inhabited atolls, making it a good environment for testing from an AEC perspective. The biggest problem with the Castle Bravo test, however, was that it far exceeded any estimates of its power and did indeed smother with fallout both Marshall Island inhabitants and a passing Japanese fishing vessel, the *Lucky Dragon 5* (misidentified as the "Fortunate Dragon" by Strauss).[68]

Strauss defended US actions and downplayed any problems, environmental or otherwise, with the Bravo test. He explained that the public and press had the wrong idea about what these Pacific atolls were like. Strauss said, "Each of these atolls is a large necklace of coral reef surrounding a lagoon two to three hundreds of square miles in area, and at various points on the reef like beads on a string appear a multitude of little islands, some a few score acres in extent—others no more than sandpits," and the US used the "small, uninhabited, treeless sand bars" for experiments. (The photos from the Crossroads test series make this statement questionable.) He further explained, "The impression that an entire atoll or even large islands have been destroyed in these tests is erroneous. It would be more accurate to say a large sandspit or reef."[69]

With his statement, Strauss also noted several other environmental phenomena worth mentioning. First, he again reinforced the importance of meteorology to tests when he discussed how, before each test shot, testers carefully surveyed the winds at all elevations up to many thousands of feet (because winds are not the same at every elevation). He also explained that testers conducted long-range weather forecasts because it takes days to do such measurements. Strauss also reluctantly admitted that, even though there was a warning area set before tests, sometimes humans did get caught in the danger zone of fallout, including the crew of the *Lucky Dragon*, the "natives," and some weather personnel. And though the tests caused some increase in "background" radiation, this decreased rapidly, and the stories about widespread contamination of tuna or other fish could not be substantiated. Instead, the only place anyone had found contaminated fish was in the hold of the *Lucky Dragon* and, of course, near the test site. These fish near the test site, though, should not have concerned anyone, according to Strauss, because "at certain seasons of the year, almost all fish caught are normally poisonous as a result of feeding on certain seasonally prevalent micro-organisms, and the natives and our Task Force personnel do not eat them at such times."[70] Whether Strauss's statement was accurate or not, it hinged upon the notion that since the only fish contaminated by radiation were not fit for human consumption anyway, they were presumably worthless by any measurement. Such thinking downplayed any environmental contamination that might have occurred because it would not have negatively affected humans. No matter the negative consequences from the 1954 hydrogen bomb tests, the United States continued to test in the Pacific.

In late 1954, the Department of Defense and AEC began to plan a deep underwater nuclear test to be held between mid-April and mid-May 1955, somewhere 200–600 miles south/southwest of San Diego, California. Robert Anderson, deputy secretary of defense, told Eisenhower in a letter that the military intended to determine "the maximum range at which hull-splitting damage to a submerged submarine at a single depth can be assured." The exact area in which the test would be conducted would "be determined more closely upon completion of special oceanographic studies now being carried out by the Scripps Institute of Oceanography and the Office of Naval Research." Either way, the general area was "essentially free of fish which are of commercial importance," so the idea was that, no matter where the test occurred in the area, it would not affect the fishing industry (not necessarily true, especially considering tuna's ecological role as a top predator and hence their ability to bioaccumulate toxins). Additionally, officials assured Eisenhower the ocean current and wind patterns in the area would reduce "the possibilities of contamination due to migration of fission products through ocean or air currents." Anderson tried to allay any worries about the spray of radioactive water over great distances, such as happened during the Crossroads Baker test, by assuring that conducting the test at a depth of around two thousand feet meant that there would be no significant water upheaval or wave formation. One problem, though, was that Mexican nationals would have to be evacuated from Guadalupe Island, about seventy-five miles away from the intended test area.[71]

As in previous tests, understandings of the natural world and earth systems proved essential both in planning for tests and for allaying concerns about the tests' potential harmful effects. And yet, Guadalupe Island was known for its endemic species and as one of the last safe havens for certain seals. Such statements might seem paradoxical, but would not have been so to AEC policymakers. To them, the natural world was important for how it could help them test nuclear weapons safely and with accurate data, improving a nuclear arsenal that would presumably protect the United States from Soviet aggression.

In March 1955, the AEC commissioners met to discuss the upcoming underwater test, named Operation Wigwam, and showed clear incorporation of environmental science into their decisions. The AEC finalized the shot as a thirty-kiloton bomb detonated at a depth of two thousand feet. Before anything else, the commissioners first discussed the geography of the test area and reviewed the conclusions of studies on the seismic effects, ocean surface effects, and the airborne, waterborne, and organic contamination that could be expected to result from the test. And even though nobody could predict exactly what would happen with the underwater detonation, somehow the commissioners were certain no component of the tests would "constitute a threat to health or safety." Importantly, Navy studies showed that the test site was "a marine desert avoided by fish and fishermen in which the ocean current drifts south and the prevailing wind is from the north." In short, the natural features of the area made the location ideal for an underwater atomic bomb detonation, and Strauss made a clear connection between oceanic and terrestrial deserts. This "marine desert," in unstated contrast to the biodiversity hotspots which were the tropical coral atolls but similar to the deserts of Nevada, served little purpose to humans unless used as a partition to sequester bombs' destructiveness from other, more useful environments. Both the "marine desert" and Nevada were effectively turned, in the words of environmental sociologist Valerie Kuletz, into "a geography of sacrifice."[72] Not to be deterred, the AEC then decided that the test would happen in May 1955, pending proper weather and ocean conditions.[73] After the test occurred, the July 1955 semiannual AEC report declared, "The test involved no health hazard to mainland or island inhabitants or consumers of fish."[74]

In general, after 1954, the issue of fallout and the radiation effects of weapons tests became more important, mostly for human health and military reasons. In part, this is because the public began to be much more concerned about radioactive fallout, particularly after the *Lucky Dragon* incident. AEC Commissioner Lewis Strauss sent a letter to President Eisenhower in March 1955 about an upcoming test that would be conducted around forty thousand feet high. Strauss included an article titled, "Atomic Blast Six Miles Up to Test New Air Defense: Nuclear Warhead for Missiles Use to Be Tried Out Soon

In Nevada." Strauss thought the article was important because it would "prevent apprehension by observers of the high-altitude test (forty-thousand feet) which will be seen for long distances." The piece told its readers that even humans standing at ground level directly underneath the blast would only receive 1/100th of a normal x-ray dose because the test would be so high up.[75] No matter the assurances many civilians still felt nervous.

In spring 1956, a reporter questioned the president about why the United States continued to research the hydrogen bomb, prompting Eisenhower to discuss the interconnectedness of environmental science and atomic weapons research. The president responded that the nation went ahead with testing "not to make a bigger 'bang,' not to cause more destruction, [but instead] to find out ways and means in which you can limit it, make it useful in the defensive purposes of shooting against a fleet of airplanes coming over, to reduce 'fall-out,' to make it more a military weapon and less one just of mass destruction." He closed by saying that the country knew how to make atomic bombs big, but that did not interest the United States anymore—making smaller bombs of reduced fallout did.[76] Reducing fallout required improving scientific knowledge, especially about the environment, and military requirements thus forged a stronger bond between scientific knowledge and weapons testing.

The argument that the United States needed to test so that it could reduce fallout was common. The July 1956 semiannual AEC report discussed Operation Redwing that had occurred a few months prior. Redwing was a full-scale test series at Eniwetok aimed "toward development of defensive weapons." The AEC planned such tests for earlier than they occurred but had postponed these tests due to unfavorable weather in the interest of safety, especially after the 1954 Operation Castle Bravo shot. Monitoring the weather, then, functioned as a safety precaution not only for the shot itself but also for control of the resulting fallout. Other tests in the Pacific Ocean also had mechanisms in place "to make measurement of radioactivity in sea water and in marine organisms." Testers sampled the water on the surface and at various depths, and also plankton and fish, with sampling extending "as far westward as radioactivity is detectable." Radioactivity sampling also occurred as land and marine biological surveys on Eniwetok and Bikini Atolls and lagoons.[77] Monitoring the environments around test sites thus served as a component of improving bombs and making them, as Eisenhower had said, "more a military weapon and less one just of mass destruction."

No matter the studies or assurances, though, many in the public remained wary of radioactive fallout. One letter to the president about harm to fishing industries from hydrogen bomb tests received a response that quoted Lewis Strauss, who claimed, "Our inspectors found no instance of radioactivity in any shipments of fish from Pacific waters."[78] Another letter asked the president to stop hydrogen bomb tests because these created strontium 90 that eventually made its way into the US milk supply. The White House chief of staff, Sherman Adams, responded to this letter by assuring that the United States would keep testing and developing weapons because these would keep the nation safe in the long run.[79] And one letter from a R. M. Tildesley caustically suggested that the president use the Nevada Proving Grounds as a vacation spot, arguing, "By setting out to make the biggest possible bomb to kill the most possible people, we seem to have scared the pants off ourselves. Now we are hoping for the age of the clean bomb."[80] Such letters probably did not carry the same sway as official White House policy advisers' recommendations but still functioned as inputs into the executive decision-making process.

A public statement by the president in October 1956 furthered the US government's public position that the nation's citizenry should not worry about nuclear fallout from tests. Eisenhower reminded the public that fallout had been a known issue since the very first atomic test at Trinity, and that the AEC had been "continuously engaged in the study of the biological effect of radiation." Reports on the subject were publicly available, and a 1956 findings of the National Academy of Sciences called biological damage from tests "essentially negligible." Moreover, The National Academy of Sciences' Committee on Meteorology determined "there was no evidence to indicate that climate has been in any way altered by past atomic and thermonuclear explosions."[81] In this case, Eisenhower implemented environmental knowledge to influence public opinion.

One series of tests, Operation Plumbbob, can serve as a final example of how significant focus on the interaction between environmental science and nuclear detonations had emerged by the end of Eisenhower's presidency. The Plumbbob test series particularly focused on fallout and the biomedical effects of tests, and a great many subprojects were explicitly concerned with improving environmental science. The project proposal claimed that the Plumbbob shots would "contribute significantly to the knowledge necessary for the improvement of our self-defense against enemy action in the event of war and the establishment of proper safeguards in peacetime applications of nuclear energy." The test series included projects, among many others, on "Radio-Ecological Aspects of Nuclear Fallout," intended to study "persistence of gross fission products" in the environment after fallout; "Biophysical Aspects of Fallout Phenomenology," which studied "the physical and chemical characteristics of fallout materials"; and an inquiry into "the physical and chemical characteristics of fallout materials," which made "fallout studies on raw agricultural products, such as exposed wheat dumps, corn and sugar cane stalks, and dried-fruit flats, to determine whether cleanup is possible, to recommend methods of protection, and to evaluate types of agricultural packaging." [82]

As in earlier tests, Plumbbob revealed a concern for how the test site environment might affect and be affected by a nuclear detonation. An April 1957 AEC commissioners meeting included discussion of a "Special Shot" for Plumbbob that would be underground. This test presented two major problems—containment of the radiation and accurate measurement of the yield. Moreover, the Nevada Test Site had geological conditions that would help ensure that the test shot did not cause an earthquake. Fears existed that if an earthquake did happen at the same time then the AEC might be blamed for the natural occurrence, but if an earthquake did take place seismic readings could determine where it originated. The commissioners also discussed what the likely effects of an underground firing would be, and "the extent of absorption of energy at a given geologic fault." They reached the conclusion that very little energy could be transferred "through a fault from one structure to another."[83]

When that underground test shot eventually occurred, it seemed to be a resounding success and demonstrated that policymakers had explicitly incorporated natural features into their tests. The Ranier Shot of Operation Plumbbob, fired on September 19, 1957, was a 1.7-kiloton test blast in a tunnel in a mesa at the Nevada Test Site. Intended to "eliminate fall-out, be independent of weather, and eliminate other offsite effects," evaluators declared after the test, "Practically all radioactive fission products were trapped in highly insoluble fused silica, indicating very little likelihood of ground water contamination." Even though three months later the test site still had elevated temperatures from the radiation (up to 194 degrees Fahrenheit), those testing thought the shot went so well "that devices 100 times as powerful as Ranier could be safely fired underground at the Site."[84] Downplaying the lasting radiation demonstrates little concern for environmental wellbeing, but the planning and results of the test show that the environment was not just the setting for the Rainier Shot, but also served as an important design feature.

Once the dust settled on Plumbbob, the AEC had detonated twenty-four nuclear devices and conducted six "safety experiments" on reducing fallout in the natural world at the Nevada Test Site, from March to October 1957. Two new testing techniques proved worthwhile—suspending bombs from balloons and detonating bombs deep underground. Using balloons prevented the atomic bomb's resulting fireball from touching the ground and this "appreciably" lowered the amount of radioactive fallout. The January 1958 semiannual AEC report claimed that of the tests that used this method, "none resulted in significant fallout in the test region." The report described how, for the underground tests, "a tunnel was dug horizontally into a mesa and at its end was bent in almost a complete circle." The testers placed a "device of known low yield" at the end of the tunnel, as this formation would seal off the main tunnel with rocks during the detonation so that no radiation might escape. The AEC declared, "The experiment's objective of containing all radiation was achieved."[85] Preventing fallout had been a primary objective for the Plumbbob tests, and when combined with the attention paid to how the shots interacted with the natural world it is clear that environmental science had come to play an integral role in how the United States and the AEC conceived of atomic bomb tests and their effects.

The underground tests can also help emphasize another seeming paradox in US nuclear testing and the environment. Even though nuclear weapons created dangerous fallout radiation, the US government continued to test nuclear weapons in order to create less fallout radiation. Thus one of the greatest benefits of underground nuclear tests was that these produced little-to-no atmospheric radioactive contamination or fallout but continued to improve environmental scientific knowledge. Policymakers in the AEC clearly privileged certain understandings of the natural world more than others, especially depending on how these did or did not support what the AEC perceived to be its mission and purpose within the US government (environmental protection certainly was not).

Despite a lawsuit trying to end nuclear testing, a hunger strike by citizen activists at the AEC headquarters, and one man claiming that nuclear testing during a full moon might cause flooding, the United States government continued to have few public reservations about its own tests.[86] The AEC claimed that weapons testing had the major objectives of "improved weapons; smaller, more efficient, and more rugged strategic, tactical and defensive weapons; development of strategic, tactical and defensive weapons with greatly reduced radioactive fallout."[87] The implication of such a statement is clear: nuclear testing had only gotten safer and produced less fallout so the public should not worry about it. But nonetheless, when asked about seeing an actual atomic bomb test, President Eisenhower replied at one press conference, "They won't allow me." After the laughter died down, he elaborated, "I have seen all the weapons, I just haven't been allowed to go to the tests."[88] The bombs may have been safe hypothetically or in public statements but were not safe enough when the president's wellbeing was on the line. Of course, the tests never were entirely safe, even with precautions. After one test series, Operation Hardtack, one memo stated, "The land area of the Bikini and Eniwetok Atolls, the water area of their lagoons, and the adjacent areas within three miles to seaward of the atolls and the overlying airspace will remain closed to vessels and aircraft which do not have specific clearance."[89] The craft that did have clearance likely focused on conducting surveys "to measure radioactivity in sea water and marine organisms."[90] Tests produced significant worldwide radioactive fallout, which increased over time as more nations detonated more bombs, leading to concern at the highest governmental levels for the environmental pollution caused by testing.

On Halloween Day, 1958, the United States, United Kingdom, and Soviet Union entered into nuclear test cessation talks in an attempt to achieve a full ban of all nuclear weapons tests by signing nations (the principal focus of chapter 3). This marked a pivotal moment in nuclear testing for the entire world, but it must be noted that the talks began only after the United States finished its large Hardtack test series. Initially planned to end in July 1958, the actual series did not end until much later due to safety concerns hinged upon weather conditions.[91] At that point, US tests ceased for a time as part of an agreement to work toward a treaty ban (these early talks never produced a signed treaty and the United States would again start testing during the Kennedy administration). But before tests stopped in October 1958, environmental science had been co-opted by policymakers into their decisions about US nuclear testing so much that decision makers frequently labeled individual test shots and sometimes entire test series with names that evoked environmental imagery, such as every test shot in the Hardtack series being named after a tree or plant.[92] Even at a metaphorical level, US policymakers incorporated the natural world into their testing plans.

Whether it was concern for how the weather might affect a test (or be affected by a test) or for the otters and geology of Amchitka Island, decision makers showed time and again that knowledge about the natural world mattered in their decision-making processes. The natural habitats of nuclear weapons ended up being places where the natural world was most conducive to creating robust scientific knowledge about those bombs and how they interacted with their surrounding environments. For example, policymakers saw little incongruity in dismissing concern for Amchitka's otters while at the same time canceling the test series because other environmental qualities—geological features—did not fit their testing requirements. Even with a frequent lack of concern for environmental welfare, the actions and decisions of policymakers reflect that they proved deeply conscious of the interconnections between environmental science and nuclear science within the context of testing. Considerations of the environment became part of the bombs themselves. Nuclear weapons came from the natural world, developed due to considerations of the natural world, and then upon detonation fully melded with the natural world. Nuclear weapons could flourish, it turns out, in many and varied habitats. Moreover, as time passed, the tests helped policymakers in both the White House and AEC develop an increasing awareness that they needed to consider environmental factors when planning and evaluating their bomb tests. It is past time that we recognize that the nexus of nuclear weapons and the natural world operated as a two-way street, with environmental science affecting atomic tests just as much or more as tests altered the natural world.

CHAPTER TWO

Fallout over Fallout

When trying to understand 1950s nuclear culture in the United States, one place to begin is the 1951 Office of Civil Defense film *Duck and Cover*, perhaps the archetypal civil defense propaganda film.[1] The film depicted an incredibly alert turtle—Bert—who always has his shell to keep him safe, along with a helmet and dapper bow tie. More to the point, *Duck and Cover* compared an atomic bomb blast to other dangers civilians confronted and understood, like fires and automobile accidents. Then, with Bert as an example, the short film told school children if a nuclear attack occurred they should duck under their desk and cover up to avoid the dangerous nuclear blast, flash, and any resulting broken glass. Toward the end of the film, the narrator proclaimed, "Duck and cover! That's the first thing to do—duck and cover. The next important thing to do after that is to stay covered until the danger is over."[2]

In an attempt to comfort school children, however, *Duck and Cover* lied when it told the youngsters to stay hidden under their desks until the danger ended. Nuclear bombs have one significant problem that distinguishes them from conventional explosives—nuclear bombs produce radioactive fallout, or countless radioactive particles that can have effects years after the actual explosion. These radioactive fallout pieces are never very big individually and frequently look like dust. And yet as humans came to know, this dust could be deadly and affect much larger areas than the initial bomb blast ever could, even spreading to cover the entire earth. While fallout cannot tear down buildings or destroy harbored ships, its effects on biological entities can nonetheless be devastating, including skin burns, cancers, and in severe cases death. Ducking and covering may have provided some protection from the initial nuclear blast, but radioactive fallout could cause problems for years after any atomic blast.

United States policymakers learned about the perils of radioactive fallout from the world's very first atomic detonation at Trinity (a physicist at Kodak film, Julian H. Webb, even detected fallout from it in Indiana), but this does not mean that those men overly concerned themselves with protecting humans from its dangers at that time.[3] The first fallout studies began with reconnaissance surveys near detonation sites of soil, flora, and fauna. In addition, the AEC used planes to trace any wind drift of radioactive clouds formed after nuclear tests. Later study of fallout would extend across the entire globe, and these early observations demonstrate that, while the Truman administration cared about fallout to some degree, it did not have nearly as sophisticated an understanding of fallout as it needed to protect the nation and its peoples.

An increasing sophistication of environmental science caused the Atomic Energy Commission (AEC) and executive branch to take fallout more seriously and desire to learn more about the nuclear menace. Both in their testing plans and the public relations arena, fallout progressively played a bigger role in executive policymaking over the Truman and Eisenhower presidencies, in large part because two nuclear tests during Eisenhower's first term— the 1953 "Harry" test shot during Operation Upshot-Knothole and the 1954 "Bravo" test shot during Operation Castle—particularly awakened US decision makers to the frightening possibilities of nuclear fallout.[4]

All the while, nuclear tests unintentionally coalesced into a massive, uncontrolled experiment on the interactions between radiation, ecosystems, and human bodies. Directly injecting radiation into human bodies to test its effects proved impossible due to ethical and scientific concerns. Consequently, while US scientists roughly understood the potential dangers of radioactive fallout, they had never conducted any specific experiments to determine its exact effects. Even as the AEC espoused concerns over and reassurances about human health, the commission's nuclear tests put radiation into ecosystems and human bodies with no real idea of exactly what that radioactivity's effects would be. This de facto experiment did not follow the scientific method and had no control group, but it still functioned very similarly to the radioisotope tracer studies that were at the same time birthing understandings of ecosystem ecology.[5] Hence at the same time US testers worked hard to increase precision in their scientific experiments about nuclear weapons, they inadvertently conducted an experiment that was at times just as dangerous as those bombs but had no formal oversight.

The tests themselves are perhaps the most tragic aspect of early US nuclear history. Nuclear engineer Arjun Makhijani described nuclear weapons production around the world as "A Readiness to Harm," and he intended a double meaning that nuclear weapons' *raison d'être* was to be launched at enemies in aggression but also that their production and testing during relative times of peace substantially harmed peoples worldwide. And the history of nuclear weapons testing does at times display, in Makhijani's words, "a disregard for public health."[6] The most obvious tragic aspect of the story is of course all of the anguish and suffering that nuclear weapons testing caused, particularly to people near test sites like the downwinders in Utah, Marshall Islanders, and Japanese fishermen aboard the *Fukuryū Maru*. But more insidious is how bureaucratic, institutional priorities combined with knowledge systems to create a setting where such malfeasance could occur. As this chapter shows, there was almost never an explicit desire to harm human bodies in tests. But carelessness can be just as deadly when public officials are unaware of what they do not know.

It is unknown exactly how many people worldwide died from any or all nuclear tests—rough estimates on a global scale are probably impossible. Economist Keith Meyers's "back-of-the-envelope" estimate of increased US mortality after tests (largely from cancer deaths due to dairy consumption after fallout) puts the number somewhere between 340,000 and 460,000 extra deaths between 1951 to 1973.[7] Such a significant range shows that Meyers's estimate is just that—an educated guess (he thinks the number is probably higher). Moreover, it would be impossible to pin many specific deaths to the tests. Just because the population's mortality level rose with the tests does not mean that it is determinable which cancer deaths were specifically linked to testing. What is perhaps most maddening among such carnage is that those testing nuclear weapons actually spent meaningful time and effort trying to determine how their weapons, including the fallout those produced, affected human bodies, particularly through environmental interaction. Seeking knowledge did not seem to change behavior very often, however. Understanding how the tragedy of nuclear fallout transpired requires getting into their mindsets and uncovering what they knew and when. In doing so, US policymakers seem less guilty of perfidy than imprudence. Fallout was necessarily going to be a significant issue for the United States, however, because of the US nuclear program's origins and focus on creating nuclear weapons.

The first steps toward establishing the United States Atomic Energy Commission began in 1939 with the discovery of uranium fission by German chemists Otto Hahn and Fritz Strassman and early attempts by US scientists to solicit funding from the federal government for nuclear research. War heightened these efforts, as is evident in the oft-cited letter from Albert Einstein to Franklin Roosevelt asking the president to look into ways "the element uranium may be turned into a new and important source of energy in the immediate future" and "to speed up the experimental work" currently being done at and funded by university laboratories.[8] Roosevelt did follow Einstein's advice for the most part and created the Manhattan Project, which was designed to produce the world's first atomic bomb. To this end, in 1943 work started on significant nuclear processing and research plants at Oak Ridge, Tennessee, and Hanford, Washington. This research produced the bombs used in the world's first nuclear bomb detonation, the Trinity test, as well as the bombings of Hiroshima and Nagasaki, Japan, in early August 1945.

After World War II ended, the United States attempted to transition its nuclear program from the wartime Manhattan Project into a postwar time of peace with the Atomic Energy Act of 1946. Senator Brien McMahon introduced the bill in late 1945 and President Truman signed it into law in August 1946 (the law went into effect in January 1947). Principally, the Atomic Energy Act (also known as the McMahon Act) sought to institute civilian control of atomic energy in the United States via establishment of the Atomic Energy Commission.[9] Even with such a mission, the AEC remained devoted to military goals and heavily influenced by the military for much of the Truman and Eisenhower presidencies, both because of shared goals and because military personnel frequently constituted part of the commission's membership.

The official AEC history of the period from 1947–1952 described how the nation transitioned from the secretive Manhattan Program, "completely isolated from the rest of American Life," into the AEC of 1952, where funding from the commission started to mesh laboratory research with a developing private industry. Over the time period, an "inexorable shift [occurred] in the commission's aims from the idealistic, hopeful anticipation of the peaceful atom to the grim realization that for reasons of national security atomic energy would have to continue to bear the image of war."[10] In testament to that assertion, US nuclear weapons research accelerated over Truman's presidency (twenty-six of the thirty-one US nuclear weapon tests from 1946–1952 occurred during 1951–1952). The US nuclear program began as the Manhattan Project with single-mindedness on war and remained focused on developing nuclear weapons on the eve of Eisenhower's election. President Eisenhower even proclaimed on the campaign trail in 1952 that the first responsibility of the AEC remained, in his mind, to "improve the atomic arsenal."[11] But manufacturing and testing nuclear weapons necessarily came with the cost of radioactive fallout.

Though planners showed concern about local radioactivity and its effects on the environs even from early nuclear blasts, they rarely altered plans based on that concern. The AEC did fund research into the breeding records of cattle exposed to radiation by the first detonation at Trinity. Policymakers further acted on this concern for how radiation affects domesticated animals by funding "surveys" in 1948 around tests at Bikini Atoll and in New Mexico. The medical scientists and biologists conducting those assessments focused on "the immediate victims in the plant and animal kingdoms" and determining which species were "highly vulnerable" or "more resistant" to radiation, particularly as part of the fascination with the effects of nuclear weapons as still newfound sources of scientific wonder. And yet, for all the concern about damages, the official AEC report to Congress on the final six months of 1948 downplayed the dangers of radiation. That report claimed, "Just as interesting as these immediate and striking effects, however, . . . [d]ata already available indicate that there are no appreciable hazards of external radiation for men or livestock at the New Mexico bomb site outside of the fenced area of several hundred acres surrounding the actual place of explosion."[12] Later inspections would challenge those conclusions, but at that time the data did not exist for fallout to concern overly those testing, even if they proved conscious of it. Moreover, testers thought they had radioactive fallout mostly under control and confined only to the predetermined bombsite. The July 1949 AEC Report also stressed that blasts did not cause much long-term radiation damage and, by mentioning that the first radiation injury occurred in 1896 from an X-ray, tried to convince readers that radiation was not a new, terrifying problem for the scientific community created only by atomic bombs.[13]

Truman-era policymakers thus frequently downplayed any concerns about fallout radiation, at times even going so far as to eschew tracking radioactive clouds produced by nuclear testing for fear of causing public panic. For example, in June 1951 Herbert Scoville, chief of the Armed Forces Special Weapons Project, counseled a civil defense aviation representative that aircraft would not be needed to track radioactive bomb clouds "in order to warn civil populations of possible radioactive hazards." Scoville claimed that, in the case of an air burst bomb, plane tracking would be "an unnecessary complication in the civil defense picture" both in the risk to which it would expose the crew and the panic its reports might create on the ground. In any detonation "where serious contamination might occur," the letter claimed that fallout would be local and then go downwind.[14] That said, Richard Miller, in *Under the Cloud*, reminded readers that everyone, no matter where they lived, faced dangers from nuclear testing. He argued, "the shadow of the atomic cloud was shared by not only [peoples and places close to tests] but by *most* cities and towns across the country. Like the soldiers maneuvering in the desert, every person alive during the 1950s and 1960s lived under the atomic cloud."[15] Everyone, in effect, lived downwind.

One of the first clues that fallout might be a more significant problem than previously believed came from environmental sampling nearly three thousand miles away from any nuclear bomb tests the United States had ever conducted. In 1951, rain samples taken in the northeastern United States showed traces of fallout from tests in Nevada from earlier that same year. This meant that fallout did not stay locally contained as previously believed. It was one thing to dump radiation over relatively unpopulated areas of Nevada, but when that fallout appeared over highly populated areas on the East Coast it was another. To policymakers centered in Washington, DC, it must have seemed that the problem "over there" was suddenly "right here." After that revelation, fallout monitoring increased with a sampling network established at the Eniwetok proving ground later that year and a mobile, two-person monitoring station 200–500 miles from the Nevada Testing Site in 1952. In conjunction with the Weather Bureau (a working relationship that would continue to strengthen), the AEC also set up over a hundred fixed monitoring stations.[16] The AEC's unplanned and uncontrolled experiment into the relationship between radioactive fallout, human health, and the natural world had begun in earnest.

In general, most of the concerns about fallout during the Truman administration relied on understandings from medical and not environmental science. For example, the AEC and the RAND Corporation engaged in Project Sunshine, a recurring research series devoted to uncovering how fallout affected human bodies.[17] The earliest research tended to center on strontium 90 (Sr^{90})—the fallout product most detrimental to affect human health—and deposits of it in human bones after atomic tests.[18] For radioactive fallout to enter human bones, it would first need to be ingested as part of the body's diet. Since Sr^{90} most closely resembled calcium (this is why it frequently ended up in human bones), the ways humans ingested calcium received close attention during Sunshine studies. AEC records on Sunshine show that, by the beginning of the Eisenhower presidency, researchers had started to be more cognizant of environmental connections as they turned their attention not only to dairy products like milk and cheese, but also the agricultural products that make up milk cows' feed like clover and oats.[19]

Hence while focus on fallout had begun to increase by the time Eisenhower succeeded Truman as president of the United States, it had not ascended to anywhere near the heights it would reach within the next decade. Previously, there had been relatively few reasons for policymakers to fear fallout as an incredibly menacing force. Finding fallout radiation in New England or Sr^{90} in agricultural products troubled many people, both experts and civilians. But tests initiated during Truman's presidency did not generate the substantial unforeseen fallout problems that would so capture the attention of the nation's citizenry during Eisenhower's first term. Less than six months after Eisenhower took the oath of office, a dramatic event forced his administration to take fallout more seriously.

During the spring and early summer of 1953, the United States conducted a test series at the Nevada Test Site called "Upshot-Knothole" that belched radiation into the atmosphere and first began to heighten policymakers' sensitivity to fallout. The May 19 test shot in that series, named "Harry" (some would later call it "Dirty Harry"), produced some of the most dramatic fallout radiation the United States has ever seen. Over the next few weeks, the AEC commissioners held a number of closed door meetings on the subject. At their May 21 meeting, they discussed the fallout from that shot and initially deemed that, because of precautions like advising townspeople downwind of the blast to remain indoors from nine o'clock until noon, no person exceeded the maximum permissible thirteen-week dose of radiation. The radiation cloud moved from the Nevada Test Site toward the St. George, Utah, area and eventually out to the Gulf of Mexico (where presumably it did not matter anymore).[20] Though they did not know it at the time, the people downwind of the Harry test shot unwittingly became involved in the AEC's uncontrolled experiment into human health and the environment.

During that meeting, the men erroneously assured themselves that likely no persons had been seriously injured from the fallout. Then they discussed the safeguards in place before all tests to minimize fallout contamination risks, showing they previously did have some idea about the dangers of radioactive fallout. As chapter 1 established, weather proved particularly important in minimizing fallout, but the commissioners pointed out that reducing local fallout risks did not always ensure the reduction of long-range fallout, and thus the two had to be balanced. Three weather conditions proved particularly important for shots at the Nevada Test Site, and testers tried never to detonate nuclear weapons when the winds 30,000–45,000 feet high blew in the direction of the St. George-Bunkersville area, "a vertical wind shear [was] present which would focus the blast on Las Vegas," or if the immediate forecast called for rain. The commissioners emphasized, "Weather forecasts, both long-range and local, are reviewed until half an hour before the shot and if, at any time, these criteria are not met, the shot is postponed." And yet, even with such precautions an element of luck persisted, as unpredicted local thunderstorms were always possible. Rain after a test could dump tremendous amounts of radiation in a localized area, and the commissioners assured themselves that small towns could be evacuated, if needed, and citizens in larger towns could be advised to stay inside. Because the above rules always had been followed, commissioners believed tests had been and should be safe.[21]

Despite assurances of safety, however, the AEC continued to encounter problems from Dirty Harry's fallout. In 1998, environmental sociologist Valerie Kuletz lambasted the development of US nuclear weapons in particular for their effects on vulnerable Native peoples who received little-to-no precautionary warnings about tests or safeguards. She decried the nuclear landscape as "too often ripened by sacrifice, for sacrifice, shrouded in secrecy, and plundered of its wealth." Kuletz further called the US West a "geography of sacrifice" that "shows how racism, militarism, and economic imperialism have combined to marginalize a people and a land that many within government and industry, consciously or not, regard as expendable."[22]

Additionally, some farmers claimed that the Upshot-Knothole test series caused livestock deaths and injuries to the survivors, allegations the AEC immediately investigated. One farmer alleged that some of his cattle died of radioactivity, but State of Nevada veterinarians determined the cause of death was malnutrition. Other nearby cattle died after drinking from a waterhole, with allegations that they had died from radioactive water, but "an analysis of the waterhole showed less than a maximum permissible concentration." What the AEC could not explain away, though, was that some livestock showed radiation burns, and perhaps as much as 10 percent of the nearby ten thousand sheep died sometime after from then undetermined causes. Many nearby farmers, therefore, wanted to get out of the area or have the AEC buy their livestock—one man even wanted the commission to buy his mining site. Unsurprisingly, this caused the AEC to worry about public relations, as some people "in the vicinity of the Nevada Proving Ground no longer had faith in the AEC." To counteract this, the AEC placed great importance on "choosing, for an objective presentation of the AEC 'case,' men who would enjoy the full confidence of the public."[23]

A week later, the AEC commissioners again held a private meeting and discussed fallout problems from Upshot-Knothole. While they began to include environmental science more significantly in their decision making, their conversation squarely focused on the liability the commission might suffer without reckoning how their nuclear experiments had altered the relationship between human health and the natural world. The commissioners considered involving specialists from agricultural colleges to help investigate animal deaths as "a matter of urgency." They also again discussed the sheep that had died near the proving grounds and confusingly reported their deaths "had not been caused by radiation; however, since the animals might have suffered some radiation injury it is possible that this was a contributing factor in their deaths." An exhaustive investigation into their deaths needed to be conducted, though, to satisfy AEC officials as to exactly what did kill the ungulates.[24] The general public remained unconvinced, however.[25] No matter how much the AEC asserted that peoples who had stayed inside were fine and that the commission's atomic bombs did not cause animal deaths, the public did not always believe the commission and with good reason.

Though it took until 1984 for a judge to render the final opinion in *Irene Allen vs. The United States* (*Allen et al. v. US Government*, 1979), a federal court eventually determined that fallout from nuclear tests caused cancer in some "downwinders," as the people downwind of the tests came to be called. Environmentalist-journalist Philip Fradkin wrote that those who suffered had been unusually patriotic and innocent, saying their biggest problem was that "they trusted. That was their downfall." For Fradkin, the most serious breach in the whole affair was that the US government, including the AEC, did not do more to warn the public about fallout dangers and how to protect their bodies. He wrote, "At one end of the scale of injustice, this breach of faith could be viewed as an act of sustained stupidity, while at the other it resembled a perfidious act carried out by a government against its own people."[26] Fradkin's perspective essentially posits that the federal government was either too stupid to protect its own citizens or too duplicitous to do so. On the other hand, that perspective is not entirely supported by the AEC commissioners regular meetings, which demonstrate thoughtful, intelligent bureaucrats who mostly put forth a good faith effort to protect US citizens. Though perhaps unlikely, at these somewhat early stages of US nuclear testing, it is at least possible that the US government simply did not know enough nuclear and

Fig. 2.1. Operation Castle "Bravo" shot, detonated March 1, 1954. This bomb, one of the first thermonuclear weapons detonated, caused the *Lucky Dragon* controversy and showered the Marshall Islanders with radioactive fallout. Image courtesy of the United States Department of Energy.

environmental science to protect its citizens. That would change over time.

Where the Harry shot of Operation Upshot-Knothole began to alert the United States public to the danger fallout radiation could have, the March 1, 1954, thermonuclear blast of Operation Castle's Bravo shot on the Bikini Atoll, as chapter 1 described, acted like a warning klaxon signaling fallout's possible dangers to the entire world. That particular Bravo shot was much larger than expected and spewed radioactive fallout over huge stretches of the Pacific Ocean. The photograph in figure 2.1 at first appears to downplay the size of the test because it is from so far away. However, on closer examination, the image reveals that the nuclear plume, because it reached a height of about twenty-five miles, actually soared well beyond the cloud cover, which would have only extended a few miles high. AEC Commissioner Lewis Strauss tried very hard to downplay any problems publicly, but at a May 1954 meeting, AEC commissioners reviewed "at length" the status from fallout problems from Pacific test operations. That review decided it was "undesirable" for the "inhabitants of Rongulap atoll [sic]" to go back to the island for a year.[27]

The commissioners decided that even though the Marshall Islanders were in "satisfactory condition," they would still need a suitable home. In addition to these islanders, a Japanese fishing vessel, the *Lucky Dragon*, also "received considerable fall-out in the test area," an event that would later help inspire the first Godzilla movie.[28] Not only did these fishermen suffer from radiation poisoning but the tests also caused worries about tuna contamination. Just as with the downwinders, these people too became part of the US government's uncontrolled experiment into radiation, human health, and the environment. The report to the commissioners noted, "Japanese anxiety about the possible consequences of contamination had been caused in part, at least, by the prospect of the cancellation of orders placed by American firms."[29] Concerns about radioactive fish contaminated by this thermonuclear blast continued to be a problem for the United States.

An exchange of letters about fallout damage from the Bravo test helps illuminate the United States' position on the subject, showing an intense desire to downplay any wrongdoing while still being incredibly concerned about national safety. Jane Nishiwaki, a woman from the United States married to a Japanese biophysicist, expressed grave concerns in a missive to the president in late April 1954. John Bugher, director of the AEC Division of Biology and Medicine, responded to her in early June. Bugher forwarded his response to Sherman Adams, assistant to the president, on that same day and described Nishiwaki as a potential communist or communist sympathizer teaching at a mission school in Osaka, Japan. Bugher declared to Adams that he had to be very careful with his response, because anything he told her might end up in communist hands. He did, however, say, "The attachment to her letter, prepared by her husband, contains valuable technical information concerning the Japanese fishing ship which information we have been unable to obtain from the Japanese authorities in Tokyo."[30] Bugher showed that issues of national security were paramount when dealing with concerns about radioactive fallout, but that the desire for knowledge meant communicating with Nishiwaki seemed like a good idea.

In Bugher's response to Nishiwaki, he tried to downplay most of her concerns about the damage of radioactive fallout from US tests and emphasized US control over its nuclear weapons and their effects. The AEC biology and medicine director alleged that some of Nishiwaki's fears were unfounded, such as the fear that radiation might kill anyone. Bugher proclaimed that no one had died, and stated, "As far as the patients who have been under American care are concerned, I can state that there is no serious permanent injury."

He claimed the Japanese patients seemed fine as well (this was false). Bugher's response to Nishiwaki also downplayed the problem of contaminated fish, claiming that only the fish of the *Fukuryū Maru* (*Lucky Dragon*) had been contaminated, and none of those fish entered ports in Japan, Hawaii, or the United States. He continued, "I understand that a few cargoes of fish in Japan were found to have detectable but hygienically insignificant amounts of contamination." Inspections found traces of radioactivity in two fish, but the levels in these "were substantially below that which would be important from a health standpoint." Furthermore, Bugher believed that only "fear rather than . . . actual radio-contamination danger" was a problem, and only then because of "the very substantial market disturbance which occurred."[31]

Bugher's most interesting assertion, however, downplayed the power of any US bombs before the power of the natural world. He wrote, "Something of a proper perspective in these matters is given by the sad news of the loss of hundreds of fishermen and dozens of ships in the recent storm off northern Japan. Impressive as these man-made nuclear detonations may be, they are dwarfed by the frequently occurring manifestations of nature." Whereas most of Bugher's letter to Nishiwaki diminished the power nature had to affect human bodies (by carrying radioactive fallout), at this point the director accentuated the natural world's power. The United States, per Bugher's assertions, may have been able to control seemingly mundane parts of the environment like radioactive fish, but humans certainly were powerless to stop a mighty ocean storm. Bugher's response to Nishiwaki therefore trod a fine line between asserting the United States' control of the situation and diminishing any power the nation (or any humans) had. He closed by reaffirming the need for nuclear tests, saying, "Finally, I am sure you will agree that devastating general war and tremendous suffering can be prevented only by keeping the free world overwhelmingly strong. To this end, personal inconvenience and some risks must at times be accepted by everyone of us."[32] What constitutes acceptable "personal inconvenience and some risks" varies from person to person, but it is worth noting that Bugher assumed much less risk from the Castle Bravo test than did Marshall Islanders and the affected Japanese fishermen.

The July 1954 Semiannual AEC Report to Congress further elaborated on the problems stemming from the recent US bomb tests, vacillating between accepting blame for the troubles and downplaying those issues. The report acknowledged that tests exposed both the Marshall Islanders and crew of the *Fukuryū Maru* to fallout radiation. But it contradicted Japanese press reports of grossly contaminated fish, claiming, "Informed scientific opinion, borne out by recent continuous monitoring by the Federal Food and Drug Administration of tuna fish coming to the west coast from the Pacific fishing grounds, and further supported by several years' results of AEC marine biological studies, provides no basis for alarm as to the consumption of tuna caught in the Pacific." Even if environmental science could be leveraged to allay some fears, the report also admitted that after nuclear tests "radioactive debris is distributed by normal air currents over large areas and with sufficiently sensitive instruments may be found to encircle the globe."[33] This meant that surely some of the radioactive fallout ended up over the continental United States and indeed many commercial fisheries worldwide. Nonetheless, concerns about radioactive fallout affecting living organisms, such as a July 1954 report from Formosa that "fish which have acquired slight radioactivity had been caught not far from the island," continued to vex the AEC and its public relations, no matter if tests found such fish "well within acceptable limits."[34]

Correspondence between Dean Rusk and President Eisenhower in early 1955 highlights that both the public and the US government took fallout radiation very seriously after the Upshot-Knothole Harry and Castle Bravo shots. The future secretary of state, then president of the Rockefeller Foundation, wrote to the president in late February to say that at a recent Rockefeller board meeting "there was an extended and sober discussion of a matter of deep concern to you and to all thoughtful men and women, namely, the effects of atomic radiation on living organisms." The Rockefeller Foundation had long supported nuclear research but thought the "development of nuclear weapons poses grave concerns which bear upon a wide range of human concerns, from the lethal effects of 'fall-out' to the new avenues which might be opened for more abundant and healthful life." The letter claimed that the nation needed more knowledge to settle these concerns, and the trustees wanted to help explore the effects of radiation on living organisms, especially "the possible danger to the genetic heritage of man himself." Rusk therefore approached Eisenhower seeking the president's approval that the National Academy of Sciences engage such research with the Rockefeller Foundation's financial support.[35]

Recognizing a good deal when he saw it, President Eisenhower responded in early March by saying he had been glad to receive the letter. While his response declared that the United States had radiation problems under control, he allowed that more research would likely be beneficial. Eisenhower emphasized that radiation problems did not come only from atomic bomb testing but also from peaceful developments of the atom, clearly implying that stopping bomb tests alone would not control every problem associated with atomic development. The president wrote that he had discussed Rusk's letter with AEC chair Lewis Strauss, who stated the commission had already budgeted $3 million a year "for its studies conducted in its own and university laboratories on various aspects of fall-out from weapon detonations, stack gases from atomic installations, the disposal of the wastes of separation processes, isotopes used in experimentation, etc." Nonetheless, Eisenhower mused, "it may well be that much more can and should be done." He thus promised to send the letter to Strauss so that the commissioner might arrange a meeting with the Rockefeller Board of Trustees "to explore further [Rusk's] very generous proposal."[36] Rusk's response thanked the president for his attention on the matter and said he would meet with Strauss on "whether there is a constructive and useful role for the National Academy of Sciences and the Rockefeller Foundation to play in this matter."[37]

The concern executive policymakers expressed about test radioactivity notwithstanding, their decisions frequently reflected a desire to prevent or mitigate the tests' political fallout more than prevent altogether the radioactive fallout, as that would have required stopping nuclear tests. To deal with fallout that had already occurred, those in power needed scientific knowledge of environments and how fallout affected those spaces. Sometimes, that knowledge merely served to allay fears, such as a June 1958 report that "no radioactivity attributable to Operation WIGWAM had been discovered by fish monitoring program on the west coast."[38] Wigwam was a test shot submerged two thousand feet to test how deep underwater blasts affected submarines. At other times, such as ecological studies on coral reefs at the "Eniwetok Marine Biological Laboratory," improved understandings simply increased knowledge about the natural world and how fallout affected it. Those coral reef studies were "on whole plant-animal populations and ecological systems in the Central Pacific island areas used in atomic test operations." Discerning the effects of radioactive fallout on these systems required understanding the reefs' ecology in its own right. Other research studied radioactivity, "natural or induced," present in Pacific seawater and marine life. After taking measurements of the area ("temperatures, current characteristics, salinity, and radioactivity" at various ocean depths), researchers found "minute traces" of radioactivity that they said did not affect the safety of eating Pacific fish.[39]

Research also attempted to determine radioactive fallout's distribution pattern throughout the world, which proved highly dependent on environmental factors. Unsurprisingly, the atmosphere and its conditions affected fallout more than anything else. As one report on atmospheric fallout from the AEC's Division of Biology and Medicine attested, the radiation could come in three types: local, tropospheric latitudinal, and stratospheric worldwide. The size of radioactive particles governed both the height these reached and the rate of fall, and size also determined which air currents would buffet those particles and thus direct how far these spread. Local fallout (more likely if the fireball from a detonation touched the ground) mostly contained larger particles, with smaller particles reaching the troposphere or sometimes stratosphere. Rain especially affected tropospheric fallout, and in this layer of the atmosphere radioactive fallout had a half-life (where half the radiation dissipated) of about three weeks. Global wind patterns meant a radioactive cloud, barring seasonal variations, could circle the earth in a month or two. Since fallout dispersed more slowly from north to south than east to west, latitudes with testing—the tropics—received more radiation from fallout than other latitudes. Stratospheric particles, on the other hand, fall extremely slowly and thus blanketed the entire earth. Only around 10 percent of what was stored in the stratosphere fell down into the troposphere each year, and once it left the stratosphere (and entered the troposphere) it would affect worldwide radiation levels.[40]

How the atmosphere distributed fallout radiation throughout the planet played a significant role in the AEC's unplanned, unrecognized experiment into radiation, the natural world, and human health. Fallout had no natural analog, and therefore researchers had no baseline to understand its progression through natural systems and the attendant effects; it was, to paraphrase historian J. R. McNeill, "something new under the sun."[41] But when research made its way to policymakers, it showed a direct connection between the natural world (especially the atmosphere) and fallout distribution. To make good decisions about radioactive fallout that would protect US peoples and interests, those in power needed to understand the environmental science of how fallout functioned in natural systems.

To gain such understandings, the AEC developed affiliations with other governmental organizations, particularly the Weather Bureau.[42] Emphasizing this point, significant meteorological research at the behest of the AEC shows an important working relationship between the US Weather Bureau and the AEC Division of Biology and Medicine.[43] As an example, the AEC used weather balloons to trace radioactive fallout, which involved meteorology and meteorologists in the research to detect and understand atmospheric

Fig. 2.2. Weather balloons used for stratospheric fallout monitoring. Balloons like these marked the confluence of US nuclear weapons advancement and the interests of US scientific advancement. NACP, RG 326, P Entry 73, Folder: Stratospheric Biology and Medicine, Photo: Balloon Photo in back of folder, Box 3. Location: 650/8/30/01.

radioactive fallout.[44] As can be seen in figure 2.2, what had previously been civilian weather balloons were co-opted as militaristic mechanisms for monitoring stratospheric fallout. Or, as figure 2.3 shows, "constant level balloons" could be used to determine radioactive fallout's trajectory. This too happened within a military context, as strategic US holdings like Guam, Hawaii, and Midway Islands are all visible, but oddly the most heavily populated downwind locations, like most of the Marshall Islands and Micronesia, are covered by clouds. Because of what is shown and what is hidden, the hand-drawn map comes across less as a weather map in full fidelity than it does using the natural world to clarify what the Pacific landscape looked like from purely a US military perspective.

The AEC's deepening relationship with the Weather Bureau showed that the AEC's experiment between fallout, environment, and human bodies could benefit other agencies as well. While one Weather Bureau representative would claim in 1959, "It appears that we do not have the oceanic fallout under adequate control," working with the AEC could help.[45] The relationship

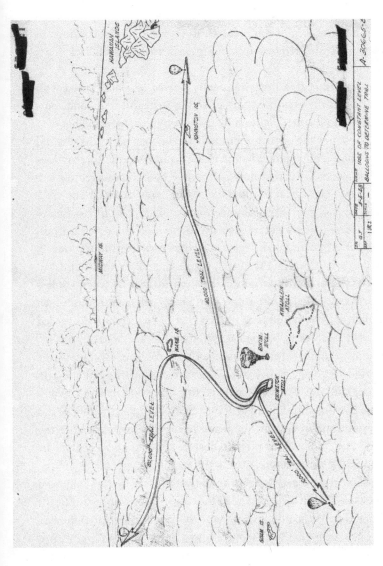

Fig. 2.3. Drawing of the use of weather balloons to monitor stratospheric fallout in the Pacific. This drawing conveniently hides most of Micronesia under cloud cover, including heavily populated downwind locales like most of the Marshall Islands. Perhaps most notably, strategic US interests—like Midway and Hawaii—feature most prominently on the map. NACP, RG 326, P Entry 73, Folder: Stratospheric Biology and Medicine, Oversized Balloon Illustration, Box 3. Location: 650/8/30/01.

benefited meteorologists, as well, since they could use fallout like a radioactive tracer to follow wind current patterns in the ozone.[46] Though testers did not intentionally do so, the fallout produced by blasts helped meteorologists run experiments on global environmental phenomena. The Weather Bureau thus became an integral part of the research to improve understandings of how fallout from testing moved throughout the earth and then fell back down to the planet's surface.

Yet decision makers at this time frequently proved less concerned with how radiation affected the environment per se but instead cared about how it might concentrate in human bodies and cause health problems. That is to say, since radiation from the environment affected bodies, policymakers did indeed care about how testing put radiation in the natural world. But that does not mean they necessarily worried how radioactive fallout damaged the environment or affected ecological health. One report explained that once fallout reached the ground, the "physical and chemical character of soils plays a predominant part in the entry of minerals into the food chain." Strontium 90, with its half-life of twenty-eight years, usually stayed in the top one to two inches of the soil. Cesium-137 (Cs^{137}) also affected environs and humans but drew less interest. Important to decision makers, though, studies found that as Sr^{90} progressed through the food chain (such as through feed into cattle and then milk into humans), the ratio of Sr^{90} to calcium decreased (Sr^{90} most modeled calcium in biological systems). Since essentially all Sr^{90} and Cs^{137} that humans ingested came as part of their diet, biological systems themselves provided a bit of a buffer or resistance to depositing radiation directly into human bodies.[47] For policymakers, this was good news—the natural world would mediate human ingestion of Sr^{90} as long as the fallout did not drop directly on human skin as happened to the downwinders after the Upshot-Knothole Harry shot or to the Japanese fishermen and Marshall Islanders after the Castle Bravo shot.

By the mid-1950s, the United States, and indeed the whole world, showed an increased focus on fallout radiation and studying it.[48] The most significant program conducted by the United States was the Department of Defense's Radiation Effects Program during Operation Redwing. That program's primary purpose "was to obtain as complete documentation as possible of fallout from high-yield thermonuclear detonations" and especially study how radioactivity distributed itself throughout the atomic cloud, collect and characterize that fallout, and correlate data so that it could be extrapolated to land surfaces. This program found that most radioactivity stayed in the lower part of the mushroom cloud and, as seems logical, larger particles created the most radioactivity. With the data collected, researchers established what seemed to be a good cloud model, "which would allow more precision in predicting areas of local fallout, although it would not be possible to predict hot spots terribly well because of variation in wind." And, lest anyone in the United States worry about the fallout created from these tests, the AEC assured that most of the Redwing test material ended up in Mexico.[49] In general, the commission tried to establish that its nuclear tests posed no threats to the safety of either government personnel or civilians. At times, efforts to assuage a worried citizenry became public prevarications, with the AEC saying it knew of no member of the public "to have suffered an overexposure to radiation as a result of living near atomic energy production or laboratory centers" or from weapons tests at the Nevada Test Site (where Upshot-Knothole Harry was detonated).[50]

Unsurprisingly, interest and focus on fallout continued to rise, and the years 1957 and 1958 saw US policymakers pay more attention to the problem of radioactive fallout than they likely had in the previous fifteen years combined. For example, in April 1957 AEC Commissioner Willard F. Libby, a University of Chicago radiochemist, reported on the "Fireball Chemistry Project." That endeavor considered "the ways and means of reducing the accessibility of radio-strontium in fallout to the biosphere and in particular to the human body. The basic technique is the incorporation of the radio-strontium in insoluble particles." In essence, this meant putting perhaps a hundred tons of sand around a test weapon before firing it so that radioactive particles might condense within that sand and trap the radiation there.[51] And if reducing radiation in tests proved important, so did reducing fears about that radiation to the public.

US decision makers on atomic matters worked hard to allay public anxieties about radioactive fallout as much as possible. For example, the AEC commissioners tried to explain in a *Parents* magazine article, "Long before nuclear weapons were even thought of, in fact, ever since people have lived on this planet, they have been subject to radiation from cosmic rays and from the radioactive material in the crust of the earth." The article claimed that, while the media might make Sr^{90} seem quite scary, the amount that had already fallen from all nuclear tests was roughly equivalent to what a person would receive in extra cosmic rays if he or she moved to a location about three hundred feet higher in elevation.[52] Such logic makes sense on one level and if the numbers were crunched in just such a way. But it glosses over the fact that the US government was still uncovering how human bodies and radiation interacted through the environment and thus did not know the full effects of nuclear tests on human health.

Public apprehension was significant enough, however, that at an early June 1957 press conference, President Eisenhower fielded a question about how "some top geneticists and other scientists have testified that fall-out radiation from nuclear weapons tests will damage hundreds of thousands and, perhaps, millions of the yet unborn in terms of physical deformities and shortened life spans." The question likely referred to something like biochemist and peace activist Linus Pauling's "An Appeal by American Scientists to the Governments and People of the World." The appeal claimed, "Each added amount of radiation causes damage to the health of human beings all over the world and causes damage to the pool of human germ plasm such as to lead to an increase in the number of seriously defective children that will be born in future generations."[53] The president brushed aside such concerns, however, referencing a previous report about how humans receive doses of radiation from all sorts of things they do every day. Diminishing the veracity of such claims, he added, "Incidentally, I noticed that [in many instances] scientists that seem to be out of their own field of competence are getting into this argument, and it looks like almost an organized affair."[54] Eisenhower insisted that the US government took its job of protecting the nation very seriously, and therefore the government had not stopped testing (even if it had tried to make those bombs cleaner). This president intended this last comment to cause the public to question its perspectives and consider which was truly more dangerous—fallout from US tests or the threat of communist hordes with their own bombs.

With public concerns about fallout rising, the White House developed a form letter to send to concerned citizens who wrote the president, which reflected a desire to moderate public opinion on the subject. The letter stated that the president certainly cared about the "possible hazard of fallout" but that he also cared about the defense of the United States. The letter even quoted former President Truman, "Let us keep our sense of proportion in the matter of radioactive fall-out. Of course, we want to keep fall-out in our tests to the absolute minimum, and we are learning to do just that. But the dangers that might occur from the fall-out in our tests involve a small sacrifice when compared to the infinitely greater evil of the use of nuclear bombs in war." Thus, the document presented the US citizenry with a hypothetical choice; "a very small risk [of fallout] from testing or . . . catastrophe which might result from a surrender of our leadership in nuclear armament which has been, we believe, the deterrent to aggression since 1945."[55] In short, the statement told those concerned about fallout in no uncertain terms that dealing with "a very small risk" of fallout from testing was the only thing standing between them and communist Ragnarok. White House staff also intended to attach, with the form letter, a copy of a report titled "Fallout from Nuclear Weapons Testing" by Charles L. Dunham, director, Division of Biology and Medicine of the AEC.

Dunham's summary report had a similar tone as the form letter and also sought to dispel fears that concerned citizens might have held about nuclear testing. The director of the AEC's biology and medicine division claimed, "No environmental hazard nor substance to which human beings are exposed has been investigated so thoroughly as radiation and radioactive materials." Dunham clearly explained that three principal dangers existed from radioactive fallout—leukemia, genetic damage to reproductive organs, and "bone irradiation" from intake of radiation such as Sr^{90}. In the end, though, the report claimed, "Pathologists agree that while theoretically there might possibly be a very small addition in the amount of bone cancer over the world as a result of assimilated strontium-90 from fallout, the effects will be so small as to be undetectable, even statistically" (not true at all, according to economist Keith Meyers's findings). Genetic effects were not really such a problem either, according to Dunham.[56] In short, while those in positions of power cared about public opinion when it came to the dangers posed by radioactive fallout, the information those decision makers put back out attempted to diminish those concerns as unfounded or convince naysayers that worrying about fallout meant a lack of concern about national security.[57] A month later when many citizens had received their AEC response letters on "the Fallout Problem," Strauss reported that he felt gratified at the surprise and pleasure many of those recipients felt. He said, "The attention given to each of the hundreds of letters received has helped to dispel some of the misconceptions held by individuals about weapons testing."[58]

Commissioner Willard F. Libby dispensed similar words of comfort during an April 1957 speech that emphasized the importance of natural factors in combating radioactive fallout from nuclear tests. The AEC reprinted the speech in its July 1957 report to Congress, assuring Libby's talk reached a wider audience. Libby explained that fallout and its effects depended on many factors, including "not only contact of the fireball with the surface, but the nature of the surface, whether it be land or water and the type of soil and the composition of the water, whether fresh or sea water." He further explained that, as a rough rule, kiloton weapons stayed in the troposphere, but megaton weapons would enter the stratosphere. After this, fallout might enter plant systems and get eaten by animals (such as cows) that would then produce milk for human consumption. Since humans receive most of their calcium from milk, this represented the most dangerous source of Sr^{90} ingestion for humans. Libby somberly noted, "Judging from experience with plants, insects, animals, and lower organisms, there is every reason to expect some genetic effects of radiation." And yet the commissioner downplayed such concerns by saying, "In summary, then, we see that the present body burden of strontium 90 from atomic weapons tests in the United States corresponds to the radiation dosage to the bones which would result from a few hundred feet increase in altitude, and the present vital statistics show no observable effect on the occurrence of bone cancer or leukemia of much larger changes in altitude."[59]

According to Libby, no matter what the gloom and doom crowd might say, the dangers from radioactive fallout once it entered the environments and biological entities humans most interacted with resulted in less radiation damage than what a person moving from the East Coast to Denver might expect to experience. By extension, anyone who worried about Sr^{90} did so unnecessarily and without proper attention to the facts. It is worth noting, however, that while other contemporary scientists raised the alarm about Sr^{90} fallout, Libby, the AEC commissioners' leading scientific voice, remained relatively unconcerned for a long period of time. He even, as historian Jacob Darwin Hamblin has explained, incorrectly believed that the planet had "infinite sinks" where Sr^{90} might be bound up long enough to decay with no harm to humans.[60]

While those in positions of power consistently tried to suppress public anxieties, they also recognized the value in improving their knowledge of the environmental science related to fallout. In the summer of 1957, AEC Commissioner Libby communicated with Eilif Dahl of the Agricultural College of Norway Botanical Institute. When asked for advice about soils and fallout, Dahl confirmed Libby's suspicion that plowing might reduce the amount of strontium that crops took in, provided it occurred in humid areas with plants that have shallower root systems (this seemed less important in arid areas where plants tend to have deeper roots). Dahl also confirmed that strontium fixation could occur in soil, meaning that the fallout would be stuck in the soil and therefore more likely to be absorbed by plants.[61]

While Libby tried to downplay the harmful effects of fallout radiation on human bodies, the AEC as an institution recognized that it needed to increase research programs in order to improve its knowledge on the connection between fallout, human bodies, and the environment. Some endeavors sought to learn more about cleaning up past messes, such as dealing with the Marshall Islanders affected by the Castle Bravo shot. By 1957, Sr^{90} had decreased sufficiently among the Marshall Islands, except among land crabs, as to allow human repopulation. The AEC figured that as long as the Marshall Islanders eliminated these land crabs from their diet and imported rice they would be fine.[62]

Returning Marshall Islanders, however, would not be fine. Anthropologists Barbara Rose Johnston and Holly M. Barker have described how island human populations suffered immensely as a result of US nuclear weapons tests, particularly the Operation Castle Bravo shot. The authors showed that, beyond immediate health effects, the Marshall Islanders experienced incredible doses of long-term radiation exposure, suffered from bioaccumulated radiation from living in irradiated environments, and encountered horrifying birth defects. Even worse, Johnson and Barker argued AEC scientists used their bodies as test subjects for medical research without the Marshall Islanders' consent, representing a significant ethical and scientific breach. It is somewhat unclear where "biomedical research . . . conducted without meaningful consent" ends and medical investigations to help a population begins, but either way it seems clear that the environments and inhabitants of the Marshall Islands were sacrificed on the altar of the US nuclear testing program.[63] As literary theorist Elizabeth DeLoughrey explained, "Overtly using the islands as laboratories and spaces of radiological experiment, British, American, and French militaries conjured those spaces deemed by Euro-American travelers as isolated and utopian into a constitutive locus of a dystopian nuclear modernity."[64]

In hindsight, the effects of radioactive fallout on the Marshall Islands functioned like the unintentional worldwide experiment in concentrated miniature. AEC investigations of course considered how radioactive fallout from US nuclear tests had directly affected Marshall Islanders, but further investigations also studied how local environments and food supply might further expose that population to radioactive risk.[65] Though not as severe as the Marshall Islanders, as radiation swirled around the world through the atmosphere, every human on the planet—especially those in the tropics—was affected. While the treatment of the Marshall Islanders is horrifying and shocking, what was intentionally done to them was unintentionally done to the rest of the world, to a lesser degree. The entire world became a subject in a similar unintentional and uncontrolled test into the effects of radioactive fallout on human bodies through their ecosystems.

By this point, the AEC fully recognized as an institution that it needed greater understanding of the natural world and its systems to help understand and mitigate the radioactive fallout produced by its bombs. Large-scale ecological studies, such as occurred in Operation Plumbbob, logically followed. Any large test series represented a significant financial undertaking for the AEC, and thus Plumbbob's approval by executive policymakers demonstrates that they understood and appreciated the importance of both fallout and the environment to understanding it. That test series began in 1957 and principally aimed to develop defensive nuclear weapons with reduced "off-site fallout," particularly using techniques such as "additional arrangements for forecasting of wind speed and directions."[66] Project 37 during Operation Plumbbob focused on the radio-ecological aspects of nuclear fallout.

Before Plumbbob, when policymakers focused on the connection between fallout and the natural world, they typically had done so in relation to human health. But Plumbbob's Project 37 research represented something new—a large research program directly into how the environment responded to radioactive fallout. Overall, Plumbbob research attempted to construct a model to determine "the manner in which [physical, chemical, and biological] parameters are influenced by variations in time, detonation yields, heights, and types of support." Particularly, Project 37 supplied researchers with massive data sets that measured the natural world and its features, which could then be used to better understand fallout. More specifically, Project 37.1 on radioecological documentation of fallout areas "centered primarily upon the relative biological accumulation, fate, and persistence of fallout products within the local flora and fauna during the acute and chronic phases of contamination." Project 37.2 "was responsible for obtaining and evaluating certain biophysical data associated with the fallout phenomenon." And Project 37.3 on agricultural soils, crops, and livestock investigated "relations within human environments and food cycles." A collection of UCLA scientists working for the AEC went so far as to clarify in a report that such "studies were dictated by a need for field data on the potential consequences to man of nuclear fallout in agricultural areas."[67] Thus even though the AEC frequently told the public it should not worry about the danger of fallout, as Eisenhower's presidency progressed the commission spent much more time, effort, and money into studying how to prevent and mitigate the radioactive byproduct of testing.

In fact, constant reassurances that problems from radioactive fallout were overblown and that the public should not worry cast a harsh light on a simple truth—the same agency that built and tested bombs was also the one charged with protecting the public from bombs' adverse effects. George Clark, a civilian geologist, pointed out this problem in an August 1957 letter to President Eisenhower. Of course, he showed his own biases by suggesting that the Geological Survey represented the ideal choice of civil servants to study the problem of radiation fallout.[68] But this letter nonetheless highlighted and explained the herky-jerky, back-and-forth nature of AEC communications, both intra-agency and to the public. The AEC spent incredible time and effort justifying the need for more tests and more information about atomic bombs and atomic energy in general. It thus makes sense that most of the facts and figures coming out of that organization (and those in the White House who received their information from the AEC) would downplay the severity of any potential atomic dangers. It is difficult, therefore, to distinguish between times when downplaying concerns reflected presenting up-to-date environmental science to the public and when it instead represented the AEC allowing its mission to improve nuclear weapons to override its goal of protecting the public from the radioactive fallout nuclear weapons tests produced.

Even internal documents, though, emphasized a perceived overblown nature of public fears about radioactive fallout. The AEC's Advisory Committee on Biology and Medicine submitted a report to the commission in October 1957 that put "The Problem" simply. It read, "The testing of nuclear weapons has injected into the atmosphere large amounts of radioactive materials in the form of dust of different particle sizes. These particles descend to the surface of the earth at different rates and constitute what is known as (radioactive) 'fallout.'" It reminded readers that both Cs^{137} and especially Sr^{90} manifested themselves in soils and milk and had long half-lives—if all weapons tests stopped at that exact moment, the report estimated that equilibrium of Sr^{90} would occur in the 1970s and decline after that. The fission products also could cause significant health issues, such as genetic damage, leukemia, bone tumors, and the obnoxiously descriptive problem of "life shortening." In the end, the advisory committee found these shortcomings somewhat inconsequential, saying, "Judging from discussion in the public press, it is not generally realized that the estimated damage is well within tolerable limits, applicable to radiation exposure of the whole population in its normal peacetime activities." Stepping out of its purview of biology and medicine, the advisory committee then claimed that the real question that needed answering was whether the size and number of bombs being tested was consistent with scientific and military requirements.[69]

No matter their continued assertions that the US populace should not worry about radioactive fallout, however, it is clear that US policymakers did indeed care about what the public thought. Back in March 1957, the United States had declared that the nation intended "to conduct nuclear tests only in such manner as will keep world radiation from rising to more than a small fraction of the levels that might be hazardous." This intention, however, meant that the United States wanted to do more testing to develop lower fall-out weapons "so that radiation hazard may be restricted to the military target. This principle was first proved in the Pacific test series of 1956." The January 1958 AEC report to Congress emphasized these ideas and also discussed other ways the United States had attempted to lower fallout of nuclear weapons, improving both bombs and public relations. Operation Plumbbob attested to the importance in improving the knowledge necessary for lowering the fallout in nuclear weapons while only marginally affecting human health, per the AEC. Its report claimed, "Measurements and calculations of possible radiation exposures to the lungs as a result of *Plumbbob* fallout showed that the highest total accumulated dose (recorded at Eureka, Nevada, with a population of approximately five hundred) was less than that to be expected from breathing for a period of 2 weeks air which contained only the amount of radioactive materials that occurs naturally."[70] US leaders kept putting more radioactive fallout into the natural world and human bodies, and therefore the US government kept having to defend its actions to US and world citizens.

Nonetheless, the AEC continued to think of its mission as first being to protect the United States and believed the greatest threat to the nation came not from radioactive contamination but from the Soviet Union. One memo from AEC commissioner Harold Vance to commission chair Lewis Strauss claimed, "In order to put the Russians on the defensive end to swing world opinion behind the United States, it is suggested that we propose an agreement to suspend the testing and manufacture of so-called 'dirty weapons' leaving the way open for testing and manufacture of so-called 'clean weapons.'" Vance used concerns about radioactivity to advocate for continued nuclear testing to create weapons with lessened fallout because "Both local and worldwide contamination from fallout would considerably reduce the fruits of a military victory."[71] Vance's somewhat convoluted logic stood in contrast to others who thought studies of radioactive fallout should be a much greater part of public policy, and the US government had to deal with both public complaints and questions about the legitimacy of its decision-making power.

In May 1958, scientist-activist Barry Commoner wrote an article in *Science* called "The Fallout Problem" that called for scientists to take a greater role in the management of public affairs because he believed the US government did not hold a proper balance of protection against communists versus fallout. Commoner saw the problem as having two thrusts—scientists did not know enough about fallout (so knowledge should be improved through further experimentation), and the public needed better access to what was known so that they could, with scientists' guidance, make better decisions. As for the science problem, Commoner explained it quite clearly, "In part, our present troubles derive from the unequal pace of the development of physics and biology. We understand nuclear energy well enough to explode great quantities of radioactive materials into the atmosphere. But our present knowledge of biology and its attendant sciences is not adequate for contending with the difficulties that follow when the radioactive dust settles back to earth." Even before the scientific process advanced biological knowledge to a sufficient level, though, the article challenged scientists to marshal "the full assemblage of facts about fallout, their meaning and uncertainties, and report them to the widest possible audience." And yet reporting to the public what was known (or believed to be known) about fallout frequently caused the public to want different policy decisions than policymakers wanted to make.[72]

Public criticisms of fallout and its effects on the environment and human bodies continued to exist and weighed heavily on the thoughts of decision makers, at least in determining how to dismiss or counter these. Alfred Phillips, a Democrat staunchly against nuclear testing, wrote a letter to President Eisenhower that asserted, "As a former member of the US Congress [D-CT] I believe and am reliably informed that everytime an atom bomb is fired it will kill 500,000 children with cancer in the blood stream. Furthermore people think that the rains, the snows and the storms can be blamed on the atom bombs disturbing the upper air. Furthermore I have information that everytime an atom bomb is fired, everywhere on Earth, the upper air streams concentrate the dangerous fallout in New England."[73] Though wrong on all three accounts, Phillips represented a section of the public that was angry and scared. Another citizen, in a letter to Eisenhower, similarly claimed, "We living today are trustees of our world and we cannot ignore the findings of science that this radio activity gradually created by nuclear tests represents a grave danger for all parts of the world, poisoning air, soil and water, affecting people, many fatally, for generations to come."[74] Representative of a common viewpoint, many people in the United States believed nuclear fallout from testing (let alone the possibility of nuclear war) severely damaged the Earth's environment, which then poisoned humans.[75] Thus US citizens worried about nuclear weapons testing despite reassuring words and releases from the AEC and even Senator Clifford Case's claims that the United States "must not let our enemies succeed in using the fear of poisoning the atmosphere—a fear felt increasingly by millions both in America and abroad—to halt our testing and development of weapons which may be essential to our very survival and to the protection of freedom everywhere on earth."[76] To give it the ammunition to allay such concerns, the AEC continued its research programs into fallout and radiation.

By mid-1958, the AEC had begun thinking about fallout in much more ecological ways and actually used that term to describe investigations "into the effects of strontium 90 on man and his environment, on the distribution, uptake, accumulation and eventual deposition in bone of strontium 90, and on methods of removing it from biological materials and from the soil." At different sites, the AEC studied how different environments circulated and dealt with the fission products, especially interested in "the long-term effects of low-level radiation on plant and animal populations." In one study at the Hanford nuclear processing plant, the AEC even created a "simulated natural pond" for the purpose of adding Cs^{137} and studying its dissipation throughout the ecosystem. Other research uncovered that nuclear test shots from towers produced much greater amounts of fallout than did balloon-supported shots (the particle size dropped from forty-four microns in diameter to five when using balloons for the tests, with smaller fallout particles containing less radiation than larger).[77] The AEC also conducted explicitly ecological studies in the Pacific, such as on Rongelap Island in the Marshall Islands.[78] Yet the commission found some studies forced upon it.

In an incident more than vaguely reminiscent of the fallout-contaminated sheep in the Upshot-Knothole Harry shot, in 1959 the AEC investigated sick cattle in South Dakota. Floyd Fishel, a farmer living near Belle Fourche, South Dakota, claimed that fifteen of his yearling calves along with a few other cattle became ill from fallout radiation. The South Dakota Division of Radiological Health of the Public Health Service took these claims seriously, as almost two years prior there had been a community-reported "radiation incident." Investigators looked into the feedlots and hay storage and eventually decided that radiation had not killed the cattle; mucosal disease had. Their findings declared that radiation did not contaminate the hay "in sufficient quantity to be responsible for the death of his cattle, "even if that hay was indeed more radioactive than usual. In the end, officials hoped this analysis would be enough to prevent "what might have been another highly publicized radiation episode."[79]

Of course, it took the AEC almost two years to finally make such judgments, meaning that any public relations damage that could have been done already likely would have been. And any persons predisposed to distrust the US government and Atomic Energy Commission on the subject likely would have found no reason to change their minds.

After the United States began test cessation talks with the Soviet Union on October 31, 1958, though nuclear tests had stopped for a time, matters of radioactive fallout and the threats it posed continued to stay in the public view. One March 1959 briefing on fallout said, "The data of the most recent [U.S. Public Health Surface] report show that SR^{90} content of milk for the month of December has decreased from the high during the period following cessation of tests, but it is possible at any time for local areas to yield a high value over and above regional variation."[80] In many ways, the ultimate measure of wheat contamination eventually would be the degree to which it affected human bones, and these high readings of radioactivity in wheat provoked anxiety. Concerns existed regarding other foodstuffs as well, and particularly high measurements of radiation in Minnesotan wheat forced the AEC to respond to public concerns.

At an early March 1959 meeting, the AEC commissioners discussed an upcoming cabinet meeting with the president on that radioactive Minnesotan wheat and the need to respond to public concerns about it. Recently released results from 1957 examinations showed the levels of radioactivity in some Minnesotan wheat had been between 105–55 sunshine units (the measurement for Sr^{90} contamination, after Project Sunshine, the recurring research series devoted to uncovering how fallout affected human bodies), which exceeded the maximum permissible level in human bones of 50–100 units. However, the amount of radiation in the wheat would in no way directly translate into the same amount in human bones if consumed, and commissioner Willard Libby reported, "If an individual ate only 'hot' wheat all of his life, he might approach the radioactive tolerance limit." Libby further explained that he thought "the lack of public understanding of radioactive fallout was an educational problem."[81] Such an assertion implied that worrying about radioactive fallout in this case represented ignorance—anyone with the proper education would have the right perspective and know that this wheat posed no real danger.

At the actual cabinet meeting with President Eisenhower, Dr. Libby talked about the same radiation issues in Minnesotan wheat, which shows that the matter was of enough importance that the president himself became involved. Libby again explained that the maximum permissible levels of radiation ranged from 50–100 sunshine units, in contrast to the present general level of just one or two units. In a moment of candor, however, the commissioner remarked that this top amount allowed was the AEC's worker tolerance level, even though the true level at which deleterious effects might be expected was still unknown. Since it was impossible to run intentional experiments on human bodies, the AEC had not done so, but Libby "concluded by estimating the hazard of radiation to be very small compared to other hazards of life." Libby thought that even though the wheat in Minnesota had been the most contaminated found, likely due to a combination of both US tests and "particularly the extremely 'dirty' tests of the Russians last October," no real danger existed.[82]

It should be pointed out, however, that Willard Libby presented a viewpoint not backed up by hard evidence and overly charitable to the AEC. Libby, a University of Chicago physical chemist, believed that the troposphere "constituted a nearly impenetrable atmospheric boundary layer" that would protect humans from radioactive fallout. Instead, as the Kodak scientist detecting fallout after the Trinity blast should have demonstrated back in 1945, fallout could have a much broader geographical range of effects and certainly was dangerous to humans.[83]

In the end, the meeting demonstrated both executive decision makers' commitment to nuclear testing and the fact that nuclear tests functioned as unplanned experiments on the nexus of human health and the environment. For example, the thought that Minnesotan wheat might become so radioactive as to require discussing it with the president likely was not previously considered by the AEC. Near the end of the meeting, the participants again emphasized that the United States needed to keep testing to develop cleaner weapons for use in any hostilities. The meeting minutes finished, "The President concluded the discussion with a comment on the difficulty of any assumption there could be a nuclear war, since the radioactivity level from a massive attack would be just tremendous compared with what is evident in Minnesotan wheat as the result merely of a few tests."[84] After nuclear tests temporarily ceased for test cessation treaty talks in 1958, though, the worldwide levels of radioactivity in food continued to drop and officials had a plan that if civilians started to worry because of any news stories they should be reminded that "Temporary rise in strontium levels in one or some foods need cause no concern; it is [the] long-term average [over an] entire diet that counts."[85] No matter the levels, though, it is clear that when the unintentional experiment conducted by the AEC affected persons outside US borders it was one thing, but when domestic health was involved it was entirely another.[86] A cabinet-level meeting with the president over contaminated domestic wheat shows the importance of geopolitical factors, and also affirms that serious problems existed when deleterious effects of testing touched the United States. As evinced by the meeting, the US government took seriously its mandate to protect US citizens in concept, even if its actions sometimes reflected other priorities.

At an April 1959 meeting, the AEC commissioners discussed whether or not the AEC should provide testimony on the scientific aspects of fallout at the upcoming fallout hearings held by the Congressional Joint Committee on Atomic Energy. Beyond scientific and moral questions about whether the AEC should participate, the commissioners had to decide whether joining the hearings would do any good at all. Commissioner Harold Vance thought that no education campaign could ever fully allay public concern about fallout. "Therefore, since all fallout to date has resulted from nuclear weapons detonations in the atmosphere, the only way to eliminate this public concern is to achieve an international agreement banning all atmospheric weapons tests."[87]

The AEC's annual reports to Congress for 1959 and 1960 contained a smorgasbord of information on fallout and reflected a very different organizational state than when Harry Truman had been in office. The AEC reported on everything from a medical reexamination of the Marshall Islanders to Sr^{90} levels in soils around the world and gave a "Chronology of Fallout Studies." An appendix after the 1959 report even focused specifically on the fallout from tests at the Nevada Test Site, discussing the approximately one megaton of fission energy released there from nuclear detonations and what happened to the 400–600 billion curies of radiation those produced.[88] The report on 1960 marked the last time that the AEC would focus as much on fallout because an executive order in August 1959 assigned the responsibility "for monitoring environmental levels of radioactivity resulting from fallout" to the Department of Health, Education, and Welfare, giving "that agency primary responsibility within the executive branch of the Federal Government for the collation, analysis, and interpretation of such data." After April 28, 1960, the AEC started giving all of its information to that department for them to publish, meaning the section in the 1960 report on "Fallout Measurements in Foods And in Man" would not be under the AEC's purview in the future.[89]

In November 1960, the AEC Office of Technical Information produced an informational pamphlet on the "Program of the United States Government in Atmospheric Radioactivity" that summed up the available knowledge and thought patterns at the time. Fallout sampling had improved dramatically over the years and involved aggregating the work of many different agencies and contractors who, in separate efforts, created a great deal of data on the subject, even though studying fallout had not been the original reason for producing the data sets. The booklet thus described a network created over two presidencies but more importantly depicted an institutional desire to learn more about radioactive fallout and how it entered and affected the natural world and human bodies. The two basic objectives of US fallout programs were understanding not only the relationship between atmospheric radioactivity input and the meteorological factors that led to the space-time models of fallout but also the relationship between surface deposition and surface air concentration to develop a model to predict distribution. As the author stated, "In summation, it is my feeling that the total level of effort will increase; that the scope will shift but that basic program objectives will not change; that complexity will increase greatly in facing the nuclear power and the space age atmospheric radioactivity problems."[90] A sea change in focus on fallout was evident by the end of Dwight Eisenhower's term, as illustrated by the Office of Technical Information's 1960 publication.

Radioactive fallout and the decisions made by those in power represent a distinct moment in US history where issues of national security, environmental knowledge, and human health converged into one single discussion about how nuclear weapons testing should occur. This confluence of factors led policymakers to consider both the environment and human bodies as integral parts of their national security decisions, as they had to balance keeping the United States safe in a geopolitical context while still safeguarding its peoples from radiation poisoning. The AEC consistently decided that protecting the nation meant continuing nuclear tests, but doing so required caring about what those tests did to the natural world. It would be easy to say that US policymakers consistently sacrificed environmental and human health for perceived safety with nuclear weapons, and often that did occur.[91] But that interpretation ignores all of the work to reduce fallout from tests and study the effects of any fallout that bombs did make. The truth is that decisions about nuclear fallout encompassed all of these factors and decision makers balanced these issues as they thought appropriate. Some of their choices had incredibly damaging effects that irrevocably harmed human bodies and extensively damaged ecosystems. But that does not mean that the environment, especially via environmental science, did not matter to US policymakers—it just means that they alternatively thought sacrificing national security was a worse option than continued testing that spewed radioactive fallout into the atmosphere, effectively turning the entire world population into subjects in an uncontrolled experiment on radiation, bodies, and the natural world.

CHAPTER THREE

Cold War Environmental Diplomacy

In 1963, several years after Dwight Eisenhower left office, a treaty emerged to curb nuclear testing, and many hoped it would eventually lead to full world-wide nuclear disarmament. Credited as one of President John F. Kennedy's real successes, the 1963 Partial Nuclear Test Ban Treaty prohibited tests in the atmosphere, outer space, and underwater—all environments where any nuclear detonation could be reliably detected from outside the testing country. The treaty represented a critical moment where, after so many hundreds of nuclear tests, international concern for human health and the environment helped lead to a political truce between hostile nations, as concern for nuclear fallout provided a major impetus for the test ban. In 1965, scientist-activist Barry Commoner called the 1963 treaty "the most important social action ever taken to conserve the quality of water, air, and the soil," dubbing radioactive fallout from nuclear weapons testing the "greatest single cause of environmental contamination of this planet."[1] Even considering the inherent intersections of environment and international relations, the role played by the natural world in forming this treaty has not been fully appreciated.

Though the treaty would eventually be signed during Kennedy's presidency, treaty talks began much earlier. This chapter demonstrates that Eisenhower-era policymakers explicitly utilized environmental science when negotiating these earlier nuclear test cessation talks. Preventing radioactive fallout helped spur the talks and became more important institutionally as negotiations progressed.[2] More concretely, knowledge about earth systems was vital in regard to detecting possible blasts. From the US perspective, a treaty was only as good as it was enforceable. That viewpoint led to an institutional position that craved environmental knowledge and sought to improve environmental science for how it could buttress US national security and foreign policy by aiding in the detection of nuclear blasts. In short, the natural world itself became a key arena of conflict in a cold war battle for nuclear supremacy as both sides sought to expand their knowledge of the natural world in order to achieve an edge in nuclear technologies.

Reducing the amount of radioactive fallout in the world was of course a goal of US policymakers. But, more than that, they perceived a need for reasonable safeguards and refused to accept any treaty that did not include provisions for detecting treaty violators. When Soviet negotiators balked at what they considered invasive and unnecessary test detection methods, US policymakers argued that only Soviet intransigence on the issue prevented a signed treaty. In effect, the United States used concern for the natural world as a negotiating strategy to improve its position and weaken Soviet bargaining power. At multiple levels, therefore, the United States leveraged environmental science to meet its aims and endeavored to improve the nation's environmental scientific understandings and measurement capabilities.[3] To consider the talks from an environmental perspective is to recognize that policymakers' conceptions of how the environment could be quantified and measured affected their ability to craft a nuclear test ban treaty. Environmental systems and scientific understandings of those must therefore be placed front-and-center in any history of 1950s nuclear test ban talks. Historical scholarship has yet to account fully for that dynamic in the Eisenhower-era nuclear test ban talks between the United States and Soviet Union.[4]

By the end of the Eisenhower administration, policymakers had fully enmeshed environmental science into their administration of nuclear technologies. Yet for all the country's technological advances, radioactive fallout forced US decision makers to reevaluate what it meant to protect the United States and its people, leading those policymakers to seek some sort of ban on testing nuclear weapons. Historian Toshihiro Higuchi has utilized a relations of definition framework to contend the Cold War created risk politics that not only escalated nuclear testing but also helped to mitigate the radiological consequences of those tests.[5] Hence incorporating environmental science into US policymaking functioned as a way to manage one set of nuclear technologies and the undesirable outcomes it produced.[6]

Because the US nuclear program had possessed such a distinct focus on weapons production to that point in its history, test cessation treaty talks represented something new and unusual. To that point, the United States had demonstrated a significant commitment to developing nuclear weapons, and the 1954 Operation Castle Bravo bomb, an unexpectedly powerful, hugely destructive thermonuclear weapon in the Pacific Ocean that manifestly altered the world's thoughts on the destructiveness of splitting the atom, provides an excellent example of that dedication to advancing the US nuclear arsenal.[7]

While previous politicians had tried to beat the atomic sword into a plowshare, Eisenhower's "Atoms for Peace" program did more to advance such plans.[8] In a speech before the United Nations in December 1953, Eisenhower asserted, "It is not enough to take this weapon out of the hands of the soldiers. It must be put into the hands of those who will know how to strip its military casing and adapt it to the arts of peace."[9] Moreover, both his work on establishing an International Atomic Energy Agency and talks about nuclear test cessation reinforce that Eisenhower was willing to consider restrictions on the development of nuclear technologies, especially nuclear weapons. Hence, even though Eisenhower devoted significant resources to improving the destructive potential of nuclear weapons, he also ardently believed in discovering and promoting the peaceful uses of atomic energy.

For all those reasons, in the fall of 1956, when Dwight Eisenhower issued a public statement on potential nuclear disarmament, he did something new. While disarmament and test cessation certainly remain separate issues, at this time many policymakers conflated the two, or at least saw test suspension as the first step toward a disarmament plan. The statement therefore laid out many of the United States' positions and concerns on the subject. Eisenhower told the nation's people he considered it "in the public interest" to give "a full and explicit review of [US] policies and actions with respect to the development and testing of nuclear weapons, as these affect our national defense, our efforts toward world disarmament, and our quest of a secure and just peace for all nations." The president stressed two tasks. First, the United States should "seek assiduously" to develop international agreements to "promote trust and understanding among all peoples." The second point, seemingly in contrast with the first point, was that the US should create nuclear weapons of both a high enough quality and quantity "to dissuade any other nation from the temptation of aggression." Eisenhower elaborated, "Thus do we develop weapons, not to wage war, but to prevent war."[10]

Speaking as much to other nations as to American citizens, Eisenhower insisted that the US would prove "unremitting" in trying to "ease the burden of armaments for all the world, to establish effective international control of the testing and use of all nuclear weapons, and to promote international use of atomic energy for the needs and purposes of peace." But the nation also insisted on establishing effective safeguards or controls in any disarmament program. Indeed, Eisenhower suggested, the only reason such a program had not yet been implemented was because the Soviet Union had not accepted any "dependable system of mutual safeguards," rejecting fourteen US proposals over the previous two years. Without Soviet cooperation, the United States had no recourse but to continue to enlarge its stockpile of nuclear weapons and needed to continue testing them.[11]

Eisenhower's statement continued by presenting a complicated duality to the public as it paradoxically contended that both the reason to stop nuclear tests and the reason to keep testing both centered on reducing nuclear fallout. The president claimed that hydrogen bomb tests were safe and did not endanger human health. Far from the truth as the Castle Bravo test had so poignantly demonstrated two years earlier, Eisenhower held to this official AEC position during the election year and insisted the United States needed to continue testing because doing so would enable scientists to reduce the dangerous fallout of future bombs the United States might use in warfare. The president continued that all nuclear bombs, no matter their size, produce fallout, and "thus, the idea that we can 'stop sending this dangerous material into the air'—by concentrating upon small fission weapons—is based upon apparent unawareness of the facts." Testing bombs would always lead to some fallout, no matter the size of weapon tested, but continued tests and experiments might produce the necessary scientific knowledge to reduce the fallout from future detonations.[12] Even though Eisenhower was dismissive of the potential danger fallout represented, reducing the environmental contamination nuclear tests produced in the form of radioactive fallout and getting the toxins out of "the air" comprised one of the few ways to make nuclear testing safe, or safer, to humans.

Despite technological limitations, Eisenhower put his finger directly on the sticking point of all negotiations to come during his term—detection systems and environmental science. He told the US people that the Soviets wanted "plans for disarmament . . . based on simple voluntary agreements. Now, as always, this formula allows for no safeguards, no control, no inspection." The president feared that simply trusting the Soviet Union might lead to serious treaty violations. If the US honored the test cessation agreement, but the USSR did not, the US could lose its lead in nuclear weapons technology. This might cause a "serious military disadvantage" for the United States if research continued without testing.[13] Either way, for test detection systems to function properly, the negotiating parties needed nuanced understandings of how nuclear bomb detonations influenced environmental processes, especially meteorology and seismology. Closely observing and measuring the earth and its systems constituted the only way to monitor for possible treaty violations. Until such an agreement could be reached for the safety of both the nation and the world, however, the president argued that the United States needed to keep developing nuclear weapons "until properly safeguarded international agreements can be reached" and do so "for the sake of our own national safety, for the sake of all free nations, for the sake of peace itself."[14] These seemingly conflicting conclusions and the logic that preceded them set up the general US position for the next several years.

Developing the US nuclear arsenal could act as a deterrent toward Soviet aggression, and improved nuclear bombs might produce less fallout, reducing the threat of radioactive fallout from future bombs; a worthy goal even though the administration downplayed the degree to which such tests polluted the planet and endangered the world populace. Fundamentally, disagreements over inspection plans were crucial from the outset to the Eisenhower administration. As President Eisenhower's prepress conference notes from May 22, 1957, instructed him if questioned about whether aerial inspection planes, necessary for "adequate inspection," might be armed, he responded, "They will be, with cameras, not with bombs."[15]

Disarmament continued to be the issue in 1957, and concerns over fallout and nuclear testing formed a crucial part of that debate. At an early June 1957 press conference, a reporter questioned President Eisenhower about how he planned to deal with the country's "anxiety" over fallout and whether he intended to modify testing plans. The President responded, "The plans that we have for testing are all bound up in the plans we have for disarmament, which we think is necessary." Eisenhower's justification was that when scientists "believe they have found something that makes [bombs] cleaner, better, more efficient" they need to test the new weapons. Later in the press conference, someone again questioned the president about disarmament and he responded that the United States could only support disarmament if it was agreed that there would be no more atomic bombs in war. He explained, "We couldn't enter into any program which forever banned tests unless we also had a system which we knew would and could be convinced would, forever ban the use of weapons, of these weapons in war."[16] Further press conferences showed an increasing focus on test cessation and radioactive fallout.

Later in the month, Eisenhower again received queries on testing and focused on testing's necessity for reducing nuclear fallout. One reporter declared that there seemed to be "a hesitancy on the part of the government to an un-equivocal yes or no on this business of immediate suspension of nuclear testing." After first responding that testing was "one of the most compli-cated subjects that the government has to deal with," the president said that the US stood by its offer to cease tests as a first step toward disarmament, as long as plans included a proper inspection system "to make certain that the whole scheme was being carried out faithfully on both sides." Moreover, Eisenhower stood by the necessity of testing to develop cleaner weapons—a "clean" nuclear weapon would produce little-to-no radioactive fallout. He told the press that the United States "had succeeded in reducing the radioactive fallout from bombs by at least 90 per cent," and that "it was certain" further testing could make that number closer to "95, 96 per cent, which is getting very close to [no fallout]." This was important for two reasons. If a bomb were going to be used for peaceful matters (like a big stick of nuclear dynamite, as in Project Plowshare), it would need to be completely clean.[17] If used in war, a clean bomb would allow the US military to confine the damage only to the desired target. Atomic bombs were destructive enough, and adding toxic ra-dioactive fallout on top of the blast destruction seemed like a bad idea to US decision makers. Of course, no guarantees existed that the Soviets would use clean bombs if they ever attacked the United States.[18]

Behind the scenes, policymakers expressed similar concerns about nuclear fallout from atomic testing. Moreover, the ambivalence in Eisenhower's posi-tion in regard to wanting to keep testing (to develop "clean bombs") in order to cease testing was reflected in other policymakers. At a meeting of the AEC commissioners that same summer, Commissioner Willard F. Libby declared that disarmament talks occurring in London reinforced the need to speed up the development of clean nuclear weapons.[19] To that point, during the fall of 1957 the AEC and Department of Defense (DOD) planned a series of nu-clear weapons tests in April 1958 at the Eniwetok Pacific Proving Grounds. The draft public announcement of those tests stated, "The United States re-peatedly has stated its willingness to suspend nuclear tests as part of a *safe-guarded* disarmament agreement. Until such an agreement is attained, *how-ever*, continued development of nuclear weapons is essential to the defense of the United States and of the Free World." The AEC intended these tests to advance clean weapons "with greatly reduced fallout" and wanted to con-duct the tests in a way that would "keep world radiation from rising to more than a small fraction of the levels that might be hazardous."[20] But no matter the push for disarmament and nuclear test cessation, some decision makers in the AEC continued to vehemently advocate the United States should continue testing.

In a memo to F. M. Dearborn, special assistant to the president, Captain John H. Morse of the AEC explained that the US needed cleaner and smaller weapons than the country already possessed, declaring, "The true justifications for further testing may be divided into considerations of *self interest*, and *moral, political,* and *military* reasons." The "self-interest" reasoning related to the dangers of nuclear fallout. If nuclear war erupted without clean weapons, the damage from that fallout would be much more significant than from any tests the US performed. Thus, even if the Soviets did not have clean weapons, the effects on worldwide radiation still were significant enough in a nuclear war that the US should only use clean weapons. Morse also believed that smaller, cleaner weapons might reduce the chance of unlimited nuclear war. He continued, "Energy serves man to the extent he tames it. Nuclear explosives are no exception," and further explained that test cessation "slams this door" shut when small nuclear bombs might be used just like dynamite.[21] The hubris in Morse's last statement is manifest. If US nuclear history to that point had demonstrated nothing else, expecting to "tame" energy was an entirely unreasonable goal. But the sentiment that the natural world only exists for human advancement has deep roots in US society, and Morse surely was not alone in his attitude.[22]

Morse continued his reasoning by melding environmental and national security concerns, deeming nuclear weapons that produced little-to-no fallout vital to protecting the United States. For his "moral" reason, he claimed, "To overkill by radioactivity and excessive yield is immoral." He also noted that a clean bomb could be made dirty if so desired, but the reverse was not true—a country with a clean stockpile would not lose the deterrence of dirty bombs but would be able to fight clean if it so chose. To support his "political" reason, Morse argued that inadvertently but indiscriminately killing with dirty bombs—very likely in the case of nuclear war—might cause friendly or neutral nations to change their position on the United States. Morse's "military" reason was that some targets, such as concrete runways or buried command centers, require larger blasts than can be achieved by conventional weapons. But unless such attacks used clean nuclear weapons, "Vast and deadly areas of nuclear contamination result," which would be bad for both any invading military force or later occupying force. Morse finished by saying that agreeing to stop tests implied that testing was harmful, and he thought this was not true at all. The United States should continue testing, as stopping would be to the "detriment of all mankind."[23] Not everyone agreed with Captain Morse's reasoning.

Robert E. Matteson, director of the White House Disarmament Staff, wrote a dissenting opinion back to Morse that questioned the captain's logic. The director countered Morse's assertion that the US needed cleaner weapons by pointing out that such an argument ignored that the weapons already were 95 percent clean—clean enough to confine much of the radioactive fallout from any nuclear attack over the Soviet Union. Moreover, Matteson was not convinced by Morse's "morality" reasoning and questioned, "Will morality have any real effect on what the Russians do or say?" As for the "military" reasoning, Matteson undercut Morse by countering that the Joint Chiefs of Staff had yet to offer any military reasons for continued testing.[24] In short, Matteson did not find Morse's arguments for why the United States needed to keep testing to be very convincing, especially if that reasoning might prevent a treaty with the Soviets. These arguments were about wartime use of nuclear weapons and not peacetime testing but nonetheless point to a developing institutional position that reflected concern for the effects radioactive fallout might have on the US geopolitical position.

A few weeks later, F. M. Dearborn, the original recipient of Captain Morse's missive, condensed the preceding correspondence into a memo to Eisenhower that argued for a policy based on balancing national security and environmental health. Dearborn asserted the world needed to understand the value of continued US nuclear testing, claiming that there were "real risks for mankind involved in test cessation." The memo continued that the nation should not accept the risks of ceasing tests lightly, and if the country did decide in the end to "acquiesce," it should try to "force maximum concessions" from the Soviets or even try to "pass the onus to Russia as the nation willing to perpetuate existing deadly risks to mankind by opposing our attempts to control radioactivity. Thus Russia might appear the threat to humanity, not the U.S." In short, Dearborn believed "test cessation [was] counter to the long-term welfare of mankind as well as national and free world security" because it posed a threat to efforts at developing cleaner bombs that reduced the amount of radioactive fallout any atomic detonation would spew into the atmosphere. He proposed Eisenhower announce before the United Nations that the United States would only produce clean weapons and only use them for self-defense, eliminate the danger of nuclear tests by testing them underground (or only testing clean versions), and share clean nuclear weapon technology with the Soviets to "protect" humankind, pending a nuclear weapons ban with inspection and disarmament. Dearborn included an attachment on "The Case of Clean Nuclear Weapons," which was very similar to Morse's original statement.[25] In the end, however as much this recommendation may have influenced Eisenhower, it did not end up becoming policy. Divided or not, the Eisenhower administration continued down a path of negotiation and worked toward a test cessation treaty with the Soviet Union.

As historian Charles Maier pointed out, Eisenhower seemed much more willing to consider arms control in the latter years of his presidency (1958 and on) than he had earlier. Advances in Soviet missile technology influenced this opinion, as they made the idea of a costless massive retaliation infeasible.[26] Moreover, the Soviets also appeared quite willing to consider talks about a test ban, even if only on terms of their choosing. After conducting its own extensive test series in early March, on March 31, 1958, the Soviet Union announced that it would unilaterally cease all nuclear bomb tests, provided the United States and its allies did the same. At this point, attempts to create some sort of nuclear test ban cessation agreement could begin in earnest. That is, test cessation talks could begin unless the United States kept testing and did not commit to the process. The choice was not easy, however, as Soviet actions had put the United States in a difficult position. While the Soviets had just finished a large set of nuclear weapons tests, the United States had its own significant test series planned for the near future. This put pressure on US policymakers to decide what they thought was most important for the nation and its nuclear program. Questions abounded about what constituted the true threat to the world—real nuclear tests (with the goal to make cleaner weapons with less nuclear fallout) or potential nuclear war (that would surely contaminate the whole world).[27]

AEC chair Lewis Strauss, for example, championed the idea that the US should eschew the Soviet test ban offer and continue with its planned tests. He generally argued that making cleaner, less environmentally damaging bombs could function as an effective political tool for the United States. At the time, the United States still planned to conduct its Hardtack I series of tests starting in April 1958, but some worried that the "international atmosphere, with the Soviet initiative for a summit meeting and a cooling off of tensions, which has aroused wide favor does not make a favorable backdrop for initiating tests." Strauss countered, "The testing of small clean weapons bordering on and blending into the conventional would serve to offset hitherto relatively successful Communist propaganda efforts to effect a distinct separation between conventional (good) and atomic (bad) weapons."[28] Strauss thought reducing the radioactive fallout produced by US weapons would take some of the venom out of the Soviets' public relations attacks, and the United States could only do so through more nuclear tests. But the subject of continued testing was still open to some debate.

Following the unilateral Soviet declaration, the United States had several possible courses of action, but the overarching debate centered on whether or not nuclear testing should continue. No matter the course of action, questions abounded. For example if the US suspended tests, how long should that suspension last? There also existed a focus on using nuclear devices for "non-weapons purposes" such as developing "'nuclear dynamite' for a variety of peaceful purposes."[29] Policymakers did seem clear, however, that they needed to take action to ensure that the United States did not fall behind the Soviet Union in the global public relations arena, because if the Soviets stopped testing but the United States did not, the USSR could claim that the United States was not serious about making peace and instead continued to poison the atmosphere.

Others questioned whether the Soviets truly were serious in their desire for peace, pointing out that the USSR's ban came at a very self-serving moment. Senator Clifford Case (R-NJ), in an opinion mirrored by many in the executive branch, observed that the Soviet Union only suspended tests, unilaterally at that, after it had completed a long nuclear weapons test series, "described as putting into the atmosphere more radioactive material than ever before." Case argued that the United States "must not let our enemies succeed in using the fear of poisoning the atmosphere—a fear felt increasingly by millions both in America and abroad—to halt our testing and development of weapons which may be essential to our very survival and to the protection of freedom everywhere on earth." Since the United States had made great strides in the production of clean weapons, Case thought US tests could continue and would not cause "a dangerous contamination of the earth's atmosphere."[30] Soviet tests were probably not as dirty as Case implied, nor United States' tests as clean. But clearly concern for the environment and world political opinion played a role in developing the United States' position even at an early stage. Protecting the nation's citizens from feared Soviet aggression clearly had to be balanced with protecting those same people from radioactive fallout from nuclear weapons.

President Eisenhower was very clear on the issue of test cessation, however, and argued that the United States would not suspend tests if he thought such a suspension would put the nation at risk. The President's preconference notes for an April 1958 press conference declared, "The United States [sic] position is that we could not consider suspension of testing as a purely propaganda measure at the risk of the security of the nation." Furthermore, the notes advised the president to stress that the United States would only stop testing when production also stopped and inspections started.[31] While the debate about the relative merits of test cessation continued, a second environmental aspect of the diplomatic negotiations emerged in the guise of test detection.

Indeed, even as Eisenhower laid out his position at that press conference, he fielded a question about the general ability to detect testing. That reporter asked whether "this problem has become more difficult in terms of negotiation" due to the notion that an increased number of detection and inspection stations were needed. Basically, for the sort of comprehensive ban the United States desired, when either side detected a potential nuclear test from a seismograph or atmospheric reading, that side would need to send in on-site inspection teams to verify whether the readings had detected a nuclear blast or merely some natural phenomenon such as an earthquake (frequently the two were difficult to distinguish from only a seismograph). The president responded that the number of inspection stations then suggested by scientists was larger than it had been previously, and that everyone needed patience to get a final answer.[32] Since the US position held paramount the necessity of more technical and environmental knowledge to design any effective international treaty, the president proposed to organize a conference in Geneva that summer on the subject.

Because of the purview of this book, this chapter overwhelmingly focuses on the deliberations of policymakers located in the United States, only bringing in outside actors and considerations when necessary to the narrative. But the Geneva Conference of Experts is important enough to deserve elaboration. Historian Kai-Henrik Barth called the conference "a defining chapter in the history of seismology" where "Seismologists from the East and West debated the strengths and limitations of seismic-detection methods and differed in their evaluations along Cold War lines."[33] The conference's conclusions and legacy are threefold. First, the conference highlighted the limitations of monitoring capabilities that might exist in any test ban agreement. Second, it showed the increasing politicization of science, as both US and Soviet scientists interpreted seismic detection capabilities along their different ideological lines, with the United States emphasizing uncertainty in detection capabilities to push for more research and more monitoring resources. And finally, the conference showed the importance of seismology, and environmental science in general, to any test ban proceedings.

When that Conference of Experts finished, it released a report in August 1958 that claimed covenanting nations would indeed be able monitor whether other signing nations faithfully followed the established guidelines through a series of on-site environmental monitoring stations.[34] The Conference explained that tests on the earth's surface up to fifty kilometers high (not deep space tests) could be monitored effectively. Upon receiving this report, President Eisenhower announced that the United States was willing to enter into nuclear test cessation negotiations with other world nuclear powers at the end of October 1958, provided those talks hinged on the establishment of an effective monitoring system. Eisenhower proclaimed the United States' willingness to keep its test ban on a year-to-year basis, as long as an effective monitoring system stayed in place and the negotiation progress continued. The president also reminded his audience that such talks were only "significant" if they constituted the first step to world disarmament and the halt of fissionable material production.[35] In late October, Eisenhower reaffirmed his desire from August to suspend nuclear tests for a year and begin negotiations on Halloween. The United States' sole condition for these talks required that the Soviet Union also suspend tests.[36] The Soviets shortly thereafter accepted the presented terms. Talks did indeed begin in Geneva on October 31, with the convening nations beginning one-year test moratoria around the same time.

The AEC commissioners met the next day and discussed the "U.S. Position on Nuclear Test Suspension Negotiations," grounding that policy on a concern for the environments in which tests could be effectively monitored. The commissioners' general sentiment held that the Soviets scored a propaganda win with recent statements and that the United States "should develop a new position which would attempt to attain agreement to permit controlled underground testing but ban all nuclear weapons tests above ground." Inspection systems remained important to the AEC, and the commissioners believed there should be a temporary ban until such a system could be developed with an indefinite ban afterward. The commissioners noted that Eisenhower's August statement "had specifically not included a reference to underground tests."[37] This nonreference was important for two reasons. The first was that underground tests were nearly impossible to monitor and the second was that the United States intended to continue nuclear tests underground, especially what they considered to be nonweapon nuclear tests (that is to say, detonate nuclear bombs intended for nonmilitary purposes, such as "atomic dynamite").

Eisenhower released a statement a few days later on the Soviet Union's continuance of nuclear weapons testing, "despite the fact that negotiations for the suspension of testing of nuclear weapons have since October 31 been under way at Geneva." The charge stated that the United States, in August, had agreed to halt testing for a year provided the Soviet Union did as well. And though the USSR had stopped testing at its arctic proving ground, the country had continued to test at another location, violating both this agreement and a United Nations General Assembly resolution that requested the negotiating parties not test during ongoing Geneva talks. "This action by the Soviet Union," the president continued, "relieves the United States from any obligation under its offer to suspend nuclear weapons tests. However, we shall continue suspension of such tests for the time being, and we understand that the United Kingdom will do likewise. We hope that the Soviet Union will also do so." The president further elucidated the US position by saying that if the Soviets did not shortly denounce testing, the United States would be "obliged to reconsider its position."[38] Clearly the Geneva talks rested on the edge of a knife blade.

In late November 1958, the AEC commissioners met specifically to discuss the nuclear test suspension talks and fleshed out the position already established. At this time, the US delegation held three basic positions: disarmament and test suspension should not be linked (because doing so might delay or scuttle a test suspension treaty), periodic evaluations should be held to determine satisfactory progress of disarmament and control system implementation, and effective control systems were necessary to make sure testing actually stopped. The delegates considered yielding on the first two positions, but some in the meeting believed the United States should not give in on any fronts. The biggest point of contention rested on the notion of a control system for monitoring—the United States wanted one and the Soviet Union did not (the fact that Soviet delegates had agreed in principle on the need for a control system during the Geneva Technical Discussions notwithstanding).[39] Environmental science and geography proved central, as only through on-site testing of air, soil, and water from specific locations could seismological readings and atmospheric monitoring be verified to make United States feel comfortable about the authenticity of any test cessation treaty.

Distrusting the USSR, the commissioners thought such a control system should be paired with the option of US entry into Soviet territory to inspect possible tests, as they deemed atmospheric monitoring alone insufficient to deter Soviet subterfuge—more on-the-ground environmental knowledge was needed to give US decision makers the confidence to enter into a treaty. In this way, who held what measurements of the natural world and from where—scientific understandings of the natural world themselves—became battlefields in this Cold War struggle. In the end, the commissioners admitted they were in a bit of a bind; since no test shots had been fired since November 3, 1958, a de facto test cessation agreement was currently in place with no control system. Moreover, the United States had to balance world opinion with legality, as public relations were crucial. The public held great fear of "off-site fallout," though Captain Morse thought the public could be educated on fallout within a two-to-three-year period. Eventually, the commissioners decided that the AEC should propose that the United States reject the recommendations of US delegates and not give in to the Soviets at all.[40] They met later that day to review the results of their earlier meeting and proposed using Christmas as an unofficial recess to negotiations.[41] The AEC commissioners held their next meeting in early December 1958, especially to discuss the test cessation talks and come to terms about how the AEC should proceed. Though the commissioners thought the AEC was not fit to decide whether the Geneva talks should be about disarmament or not, they did believe that if disarmament was off the table the AEC should proceed with "armament development through testing, and testing would mean the need for controls."[42]

The same commissioners meeting continued with a discussion of a test cessation proposal by Senator Albert Gore Sr. (D-TN) that focused on protecting the earth from radioactive pollution produced by nuclear bomb tests. Gore recommended that Eisenhower "announce the unconditional and unilateral cessation of all nuclear tests in the earth's atmosphere for a specific period, possibly three years, and ask [for] similar action by other nuclear powers." This would keep treaty talks going and eventually might lead to discussions on the discontinuance of other types of tests. Gore's proposal stemmed specifically from being "deeply concerned with the apparent impasse in the Geneva talks and with what he believed to be an increasingly high contamination of the atmosphere with radioactivity." Commissioner Willard F. Libby countered that the Soviet tests in October had raised strontium contamination in the Earth's atmosphere by twice what US tests had during the previous four years. Since "the world cannot tolerate unlimited dissemination of airborne radioactivity," Libby believed "the President [ought to] make a public announcement concerning the amount of atmospheric radioactivity created during October by the Russians."[43] This sentiment shows that radioactive fall-out in the atmosphere mattered both for how it damaged human health and the environment and for how it could be used by both sides to score politi-cal victories with the public.

The commissioners saw several different positions the United States could take concerning Senator Gore's proposition that would lead to both political and environmental gains. The AEC could have supported Gore's proposal or merely kept talks going at their present rate while ramping up US aggressive-ness at the negotiating table. Another possibility involved sharing a technical paper with the Soviets that highlighted the costs of detecting underground tests (in contrast to relatively effective and existing means of detecting atmo-spheric tests) to prod the Soviets into agreeing to an atmospheric ban. The last choice called for Eisenhower to announce the airborne radiation previ-ous Soviet tests produced and try to publicly shame the USSR into a bilat-eral agreement to stop further atmospheric contamination. AEC Chairman McCone said that each of the latter choices could be interpreted as an at-tempt by the United States to break up the conference. Instead, he thought the United States should maintain its position that monitoring and controls were crucial for any agreement. Commissioner Harold S. Vance pointed out that the AEC had, the year previously, proposed to the president a plan sim-ilar to Gore's, which Eisenhower had rejected. Moreover, one staff member believed that any announcement similar to what Senator Gore desired should be withheld until it was clear that the Soviet delegates would not accept any control agreement.[44]

Test cessation treaty negotiations thus show that while the AEC seriously considered whether nuclear tests poisoned the planet and the human race, the idea that held the greatest sway among the commissioners was how the United States could leverage environmental concerns for political gains. Such thinking valued, at different times, both a clean planet and a polluted one. In terms of US actions, an unpolluted natural world meant healthier, better protected US citizens. And though radiation from Soviet tests might endan-ger the global populace, it could also be used at the negotiating table to wrest concessions from USSR delegates. Either way, the discussion shows a prior-itization of environmental science by officials trying to balance their respon-sibilities to the nation.

The meeting, as a whole, showed a melding of concerns about environ-ment, national security, and politics, as decisions about one affected the oth-ers. As the discussion and meeting wound down, Brigadier General A. D.

Starbird, director of military application of the AEC, claimed "that the matters under discussion were highly sensitive in that if the Russians learned of the debating going on within U.S. Government agencies regarding the U.S. position at Geneva, they would be aware of a weakening in the U.S.–U.K. position and could exploit the situation."[45] Distrust and paranoia aside, this meeting perfectly outlined the AEC's place in the negotiations. The commission had no formal power in the talks, other than its advisory role on nuclear policy to the president. It is important to remember, however, that nuclear test cessation was not necessarily in the best interests of the AEC as an organization. And yet, the AEC was not entirely against a nuclear test ban, even though a ban would have limited the agency's power. Instead, the commission balanced concerns about how tests affected the environment and human health within an institutional framework obligated to develop the nuclear weapons deemed necessary for the security of the United States.[46]

A week later, the commissioners again met to discuss Senator Gore's test suspension proposal, particularly focusing on how detection capabilities differed vastly depending on environment. Following Gore's call for the Geneva conference to proceed immediately to negotiate for a permanent stoppage of atmospheric tests, Commissioner Libby pointed out that the number of stations required to detect atmospheric tests would be much fewer than those to detect underground tests. Atmospheric-only test detection stations could also be confined only to the Soviet Union, negating any possible issues of location in China, and would thus be more economical and geopolitically feasible. However, Chairman McCone chastised the AEC report made on Gore's proposal, as he felt the report only sought to discredit Gore's ideas and provided insufficient analysis. Moreover, Commissioner Vance reminded everyone at the meeting that with both President Eisenhower's August announcement on halting US weapons tests and the Geneva conference the United States hoped to achieve complete cessation of all nuclear tests. If the United States kept the negotiations from achieving that goal, the USSR might gain a propaganda advantage by stating that the United States had never been sincere about achieving a total ban. Chairman McCone agreed that the United States should pursue the original goal of a complete ban, and, if a more elaborate detection system was later found necessary, such control systems could be considered at such a time.[47] Even if a partial ban might have been more feasible in the moment and eventually have led to a complete ban, there clearly existed a significant bloc in the AEC's leadership that thought the United States should pursue either a total ban or no ban at all.

The issue of a control system continued to vex negotiators, especially because detection capabilities differed in each testing environment. Issues of test detection existed even from the beginning of the atomic age, and the United States had endeavored to improve its test detection capabilities as far back as Truman's presidency. For example, one significant part of Operation Sandstone (April–May 1948) centered on detecting atomic detonations in the air. Because of those experiments, by September 1948 the AEC expressed confidence in "being able to detect by radiological means an atomic air-burst."[48] Moreover, the US government prized that knowledge a great deal, with considerable debate about whether information on detection capabilities should be shared with other governments, even allies such as the United Kingdom or Canada.[49]

The very first Soviet test had heightened the importance of those detection capabilities. On September 3, 1949, the United States began detecting "the emission of large quantities of radioactive material" and later determined through various systems that this came from the first Soviet atomic test.[50] Alarmed AEC commissioners wondered during a December 1949 meeting whether "all the Commission technical personnel, including the most able people, were available for this important program and were being used."[51] A year later, the United States public learned exactly how this occurred. On New Year's Eve of 1950, brothers Joseph and Stewart Alsop (two men very unpopular with the Truman administration for not toeing the official line at all times) wrote an article for the *Washington Post* titled "How Red A-Blast Was Detected." The article rankled many, especially Gordon Dean, then AEC chair, who declared that detection methods represented "a tightly held secret" that he did not appreciate the Alsop brothers divulging. President Truman himself called the article "a matter of deep concern [because] of its disclosure of highly classified information of major significance to our national security."[52] National security or not, the *Post* article clearly articulated the US method for detecting blasts.

The US detection system had four parts that depended on interpreting scientific knowledge about earth systems within a Cold War context: seismic readings, using Geiger counters to detect the presence of radioactive particles in the stratosphere, sampling air particles from the radioactive cloud, and finally having scientists interpret all the data. The Alsops explained, "When the Soviet bomb exploded in central Siberia in September, 1949, all this elaborate organization, already trial-tested by our own Eniwetok bomb, went smoothly to work." Seismologists located the explosion location and Geiger counters detected the radioactive cloud, with air samples providing the "decisive evidence" of the bomb detonation occurring and its type, as different nuclear processes (e.g., uranium bomb vs. plutonium bomb vs. an atomic pile explosion) produced very different sorts of radioactive particles.[53]

Yet for all the desired secrecy on test detection systems during the Truman presidency, by the summer of 1958 things had changed a great deal. The July 1958 semiannual AEC report to Congress contained an appendix titled, "Information on Detonations Released for Use in Seismological Studies." Though it left out some important information, such as power of the tests, the appendix released exactly when and where US tests happened and what type of test occurred so that seismologists could cross-reference their records with the AEC data.[54] Released at the same time as the Conference of Experts occurring that same summer, the AEC surely intended the data to further test cessation talks and make discussions of a monitoring network more feasible. Such an appendix especially demonstrated that the AEC was serious enough about test cessation talks as to give away important environmental knowledge to improve detection capabilities.

The AEC commissioners met again in late December 1958 and discussed the proposed AEC position on test cessation. Unsurprisingly, the issue of a control system loomed large over the proceedings. With the Geneva talks on a break surrounding Christmas, the AEC revised its official position for when talks resumed in January 1959 so that the AEC would suggest no arrangements be made on a control system until its scope could be determined. The commission would also stand by the idea that fixed detection posts would be inadequate because the Soviets might just find a way to work around any immoveable posts. The issue of whether stations were needed in China remained tricky, and General Starbird opined that he felt not including posts there would create a possible loophole. The issue elicited a lengthy discussion but no clear answers.[55] What the commissioners' deliberations made clear, however, was that while knowledge of the environment allowed for test detection under certain circumstances, similar knowledge also aided any country that desired to test nuclear weapons without the knowledge of rival nations. Whether a country desired to construct a control system or evade it, its understandings and measurements of interrelated earth systems proved vital in such efforts.

The next day, Harold Vance, as acting AEC chair, sent a memo to summarize the AEC position on detection systems to Under Secretary of State Christian A. Herter. Vance told Herter that seismological results from the underground nuclear tests conducted in Nevada during October 1958 showed that the Conference of Experts' conclusions on detection "require re-examination." In short, the detection capabilities assumed during the conference were not as robust as initially believed. The AEC therefore proposed the United States adopt a new position because of three reasons. First, the control system proposed by the Conference of Experts to detect underground and space explosions had a much more limited capability than initially believed. Second, tests in those environments did not cause fallout like atmospheric tests, meaning preventing tests underground and in outer space proved less pressing to world concerns than atmospheric tests. Finally, adequate control systems already existed for atmospheric testing and thus the United States would not have to sacrifice its core stance on detection to ban atmospheric tests.[56]

The discrepancies in detection capabilities based on environment clearly evince the differences in different forms of environmental knowledge and human abilities to monitor earth systems at the time. Atmospheric tests produced telltale radioactive particles that could be detected, measured, and compared to knowledge about atmospheric circulation to determine with some reasonable certainty the type and quality of test. Underground tests, however, produced tremors in the ground that could be picked up with seismological instruments. But determining with confidence whether an earthquake was natural or nuclear was much more difficult unless an on-site visit could be undertaken. Outer space explosions produced no earthquakes, and, being so high up, the atmospheric radiation was much harder to pin down to a specific location.

The new proposed AEC position was that the United States should adhere to the idea that only detectable and identifiable tests should be stopped, and negotiate, as a first step, a treaty for the cessation of atmospheric tests. In addition, the United States should postpone treaties about underground and space tests while simultaneously proposing international cooperation in investigating identification problems and preserving the right to develop "non-military applications of nuclear explosives."[57] The implication was that tests might eventually be banned in all environs, but only when detection measures and their related environmental knowledge advanced enough so as to be deemed adequate by the United States.

This official AEC suggestion reinforced the importance of developing robust environmental knowledge and measuring capacities in multiple environs to any test cessation agreement. Moreover, the commission emphasized that stopping atmospheric tests should be the primary goal anyway, as those tests caused radioactive fallout that damaged the environment and human health. The natural world, long before negotiators even had decided what the official goal of their talks should be, played a crucial role. Not only did concern for environmental health provide the impetus for the talks but examining the natural world (both the atmosphere and with seismological readings) was the only way to tell if a nuclear test had occurred. The state of environmental science essentially determined the US position.

Yet, by 1959, the AEC still had not settled on an official position, as differing goals between US officials made it difficult to achieve consensus about how to put existing knowledge of earth systems into practical use. In January 1959, Chairman McCone met with the other commissioners to firm up the AEC's position on test cessation so that McCone might be prepared for later meetings with the secretary of state and Department of Defense. Fundamentally, the issue of a detection system dictated the terms of those discussions. The commissioners—those with military and civilian backgrounds alike—all agreed that the AEC needed more underground testing to gain the necessary seismological knowledge to develop an adequate detection system in those environments. Whereas previous tests were deemed necessary to develop cleaner bombs, as politics shifted so did the reasons to test. Commissioner John S. Graham suggested the United States "develop an orderly program for obtaining the additional information needed for establishment of an adequate detection program" and should do so immediately, not reserved as a fallback position. He also believed all atmospheric tests should stop. Commissioner Vance pointed out that top US scientists agreed that the inspection system approved during the summer 1958 Geneva Conference of Experts was inadequate, and agreed with Graham's position that the United States needed to conduct more underground tests "to know with certainty that a detection system would be dependable," even though the earliest time more underground tests could be conducted was summer 1959.[58]

Practically, though, the Geneva talks could not wait on further research so the AEC needed to take existing environmental science and form a new base position immediately. The AEC operated within an odd space in terms of the test cessation talks. While they were not officially in charge of the talks, the commission's expertise mattered. And the AEC was still in charge of much of the nation's nuclear program and still had to administrate those technologies. After further discussion, the commissioners decided on several recommendations. First, they thought the US proposal should seek agreement among nuclear powers to suspend atmospheric tests and create "an adequate detection system to insure compliance. This step would eliminate the fall-out issue." Second, the US proposal should permit testing in outer space and underground "pending the conclusion of a series of tests in diverse geographical and geological environments and in outer space from which could be designed an effective system to detect and identify such explosions." Such tests could be joint efforts or unilateral, and the data would be shared either way.[59] More discussions continued on how measuring the earth could help detect nuclear weapons tests.

By the end of March 1959, the Panel on Seismic Improvement, sponsored by the US Department of State, reported its findings. Chaired by Lloyd Berkner, a physicist who helped organize the 1957–1958 International Geophysical Year, the panel had been charged with evaluating how the seismic detection capabilities outlined by the previous summer's Conference of Experts might be improved. Meeting several times over the early months of 1959, Berkner pushed the seismologist participants to, in the words of historian Kai-Henrik Barth, "think big. What kinds of projects would they like to pursue if increased funding would become available?"[60] That the Panel even came into existence is a testament to the need and desire for improved environmental knowledge, especially in seismology to aid in the improvement of underground detection systems.

The Berkner Panel Report's message can be summed up by its title, "The Need for Fundamental Research in Seismology." The panel explained that both earthquakes and nuclear detonations produce seismic waves, and therefore determining the difference between the two from a seismograph represented the key issue. Such a capability, however, would require greater knowledge about seismic waves. The panel concluded, "The strategic requirements of detection, together with the need to maintain a competitive position in one of the most significant fields in the earth sciences make it a matter of urgency to institute a high level of support of seismological research." The report also laid out exactly the sort of research the panel thought should be conducted.[61]

Decision makers had created the panel to consider "the feasibility of improving the capability of the system recommended by the Geneva Conference of Experts [the previous] summer to detect and identify underground events." To do so, the panel looked at ways to improve the Geneva system with existing technology, improve the system through research in seismology, and the possibility that any detection system might be circumvented via concealment of underground tests. The panel also confirmed that earlier detection methods were not as effective as previously believed and validated concerns about the difficulty in distinguishing from seismic readings alone whether a tremor had been from an earthquake or a nuclear detonation. Improved equipment and techniques would greatly improve detection capabilities for underground shots, but detection still would be limited to blasts no smaller than five-kiloton weapons. And yet, these conclusions came from still very limited data, and "the Panel concluded that a vigorous research program in seismology would result in important improvements in the ability to detect and identify earthquakes of small magnitude." Augmenting the Geneva net with unmanned stations, it concluded, might improve capabilities.[62]

Portions of the Panel's conclusions could have been anticipated—if you tell a group of seismologists to dream big and ask them what would help fix a problem, it is unsurprising that their answer might center around improving seismology. But some of their larger points were undeniable: without both environmental measurements in a variety of earth systems and improved knowledge to help interpret such readings, creating an effective control system would be all but impossible. Without that control system, the United States never would agree to any test cessation treaty.

Two weeks after the Panel on Seismic Improvement released its report, President Eisenhower wrote a public letter to Soviet Premier Nikita Khrushchev on the subject of the test cessation talks. Unsurprisingly, the issue of a detection system and how it might influence the eventual treaty took up much of the ink. Eisenhower declared that the United States "strongly seeks a lasting agreement" about test cessation, but that the agreement should be "subject to fully effective safeguards" and the "present proposals [by] the Soviet Union fall short of providing assurance of the type of effective control in which all parties can have confidence: therefore, no basis for agreement is now in sight." And yet, the president asserted that the cessation talks "must not be permitted completely to fail." He suggested talks should therefore start only with an atmospheric ban, because such tests could be monitored and would not require on-site inspections. Continuing the use of fears about radioactive fallout for political gain, Eisenhower asserted that some sort of test ban was vital to calm the public about atmospheric radiation. Doing so would cause "fears of unrestricted resumption of nuclear weapons testing with attendant additions to levels of radioactivity [to] be allayed, and we would be gaining practical experience and confidence in the operation of an international control system."[63]

Ten days later, Khrushchev responded to Eisenhower, also in a public letter, and questioned some of Eisenhower's basic assumptions about testing and the natural world. Khrushchev said he was glad that Eisenhower also was "of the opinion that these negotiations must not be allowed to fail." Yet, Khrushchev also leveraged the natural world for political gain when he argued that stopping atmospheric explosions up to fifty kilometers did not "solve the problem," nor did it prevent the production and improvement of other types of nuclear weapons. He continued, "Explosions of nuclear weapons at altitudes of more than 50 kilometers would also poison the atmosphere and the earth, contaminating with radioactive fallout the vegetation which finds its way into the food of animals and into the human organism, just as is occurring at the present time." Even though detection capabilities might have stopped at 50 kilometers, detonations at both 40 and 60 kilometers similarly affect the atmosphere and human health, meaning that a ban on tests only 50 kilometers and below represented a "dishonest deal." The Soviet premier thought that such issues should not deter either side, and the talks must find a way to cease tests of all types of nuclear weapons. Khrushchev acknowledged that the "most serious difference" between the United States and Soviet positions was the issue of inspection teams, but that no matter what the talks decided, the Soviets would "make every effort to achieve agreement on the cessation of nuclear weapons tests," telling Eisenhower that he could "be certain that if we sign a document we shall, even if there is no control whatsoever, faithfully carry out the obligations assumed by us, because for the Soviet Union public opinion and the opinion of nations is dearer than anything else."[64] No matter that both sides were attuned to environmental understandings and espoused platitudes about how tests "poison the atmosphere and the earth," differences remained unresolved.

The two public letters make it plain that the issue of test detection remained the stickiest wicket in the treaty talks. The AEC commissioners discussed the Geneva conference at two separate mid-May meetings. In the second meeting, the commissioners learned that talks would adjourn until the second week of June and that the Soviets had agreed to a proposed control system that would use satellites to observe nuclear tests in outer space.[65] After talks resumed, a reporter questioned President Eisenhower about the idea of "decoupling," or using specific techniques in underground atomic tests so that tests appeared much smaller than they actually were. For example, the seismic readings from a 10-kiloton blast might measure as the equivalent readings from a 1-kiloton test. The president responded that while concealment methods had improved, so had detection techniques, and therefore detection capabilities were roughly the same as earlier, despite technological improvements.[66]

Over that same summer, AEC chair John McCone went to Geneva to attend the Conference on Nuclear Test Cessation, and his time there led him to believe the issue of a detection system mired the talks to an even greater degree than he had previously supposed. In early July, he gave the AEC commissioners a report on his trip. McCone described a dinner he attended "as a pleasant and convivial affair at which there was a frank exchange of views among those present," but, in spite of the amiable conversations, problems still existed. For starters, the Soviets insisted that they would not agree on a specific quota about on-site inspections until both sides reached an agreement "on the idea of a quota." McCone feared that there was pressure to reach an agreement on a quota with no reference to the technical capability of the supposed inspection system, and he was afraid that if that happened public opinion might cause the United States to accept an inadequate number, such as twenty-five on-site inspections per year or fewer. Moreover, while the United States did not desire a ban on anything that could not be monitored, the Soviets still pushed for a complete ban.[67] The fact that the two sides could not even agree "on the idea of a quota" might seem a bit shocking considering how long the talks had been ongoing, but the mingling of international politics and environment has rarely produced easy answers.

Prompted by these fundamental differences between United States and USSR positions, McCone said the congressional consensus recently expressed to him had been that the conference should be recessed so that "senior Government people capable of making an objective appraisal" could reexamine the US position. Only then could the United States develop a firm stance. Upon the conference's reconvention, US delegates should state this position (especially on basic issues of inspection, staff access, and veto) and force an agreement before returning to fringe issues. And yet, US scientific advisers could not agree upon the technical components of some of the principal matters. In June, Chairman McCone had noted a planned scientific panel to evaluate test detection systems. This "Interdepartmental Panel on Test Detection" contained representatives not only from the AEC and White House but also the DOD, Central Intelligence Agency, and the State Department. But as of early July, there existed no set agreement by the Interdepartmental Panel on the number of required inspections for an adequate control system, and General Starbird contended that even more important than technical requirements was the unknown role that US intelligence reports might play in verifying any test ban violation.[68]

In August 1959, President Eisenhower extended the preexisting one-year nuclear test suspension through the year in an attempt to salvage the talks, however unlikely a test ban became as time passed.[69] This prompted the Soviet Union to state that it would not resume its own nuclear tests so long as the United States and its allies maintained their moratorium, but by December, it became clear that an agreement would not be reached before the New Year. Eisenhower could have extended the moratorium past January 1960, but some US policymakers thought the United States would be better off testing instead. In October, General Starbird wrote a memo that the US should plan to begin testing underground as soon as possible after the January 1 deadline.[70] In a December meeting, AEC Chairman McCone announced that the Congressional Joint Committee on Atomic Energy (JCAE) supported the AEC's position on the necessity of effective monitoring in any test ban agreement. Senator Clinton P. Anderson (D-NM), chair of the JCAE, "urged" that the AEC be in a position to test nuclear devices as soon as the test moratorium ended.[71] Those who wanted the United States to continue testing its ever-advancing nuclear arsenal were winning out.

Less than a week later, the commissioners met again and discussed the idea of a "phased suspension treaty" where the test ban might initially extend up into the atmosphere to 300,000 kilometers (over 186,000 miles). A ban to this height should have eliminated any fallout problems and could be monitored with a combination of both ground controls and satellite systems.[72] Reducing atmospheric pollution therefore remained important, but geopolitical strategic considerations trumped environmental and health concerns in the long run. An underground test ban remained off the table for the United States, and in December in Louisiana the AEC performed two high explosive nonnuclear detonations as part of an experiment on decoupling atomic tests (making underground blasts appear seismologically much smaller than they are).[73]

In February 1960, President Eisenhower released a statement on the test cessation negotiations. Even though he declared that the United States would start pushing a proposal to end the negotiation deadlock, his statement accentuated US intransigence on the issue of a detection system. The president reminded the public that the United States "had stood, throughout, for complete abolition of weapons testing subject only to the attainment of agreed and adequate methods of inspection and control." Following this position, the US proposal would have sought to end tests "in all the environments that can now be effectively controlled"—tests in the atmosphere, oceans, space, and underground where it could be monitored. While not a complete test ban, the proposal contained "initial, far-reaching, but readily attainable steps" that would "allay world-wide concern over possible increases in levels of . . . radioactivity into the atmosphere."[74] Such a treaty proposal looked remarkably similar to the eventual 1963 Limited Test Ban Treaty and highlighted the fundamentally environmental nature of any test cessation treaty (reducing radioactive fallout from the atmosphere and navigating the differing environmental responses produced by nuclear testing).

Several days later, expecting questions on the new US position, a briefing paper for a mid-February press conference gave the president some prepared answers that unsurprisingly cast the United States in the best possible light and concomitantly sought to put all blame for unsuccessful treaty talks on USSR diplomats. The Soviets had turned down the US proposal and countered with their own plan that would ban all nuclear detonations but only allow a limited number of inspections on suspected tests. The briefing sheet told Eisenhower that the correct "answer" was that such a policy "casts considerable doubt on the Soviet Union's professed desire to halt nuclear weapons tests," and that the Soviets are doing a "disservice" to "the hopes of all peace loving people."[75] When asked about this in the actual press conference, the president reviewed the situation and then stated that he thought the US proposal had been a good one, and that the Soviets' counterproposal seemed to change the criteria they were willing to observe. Hence deciding on an appropriate number of inspections would be difficult.[76] Even when the two sides seemed closer to reaching an agreement, fundamental differences in positions remained.

The issue of detection systems continued to vex negotiations as furthering environmental science, especially seismology, both allowed for greater test detection and highlighted the ever-increasing difficulty of detecting the covert tests of any potential treaty violators. Scientific research was thus a cat-and-mouse game of detection and evasion where the natural world itself was a battleground for nuclear supremacy. The briefing paper for one of Eisenhower's late April 1960 press conferences informed him that scientists claimed a need for 180 more testing stations in order to ensure effective monitoring. This would help the United States achieve a system to monitor a ban on "all nuclear weapons tests in the atmosphere, the oceans, at high altitudes and above seismic magnitude 4.75 in the underground area."[77] At the actual press conference, a reporter questioned the president about testimony the previous week before the JCAE on how "the art of concealing underground tests was outstripping the art of detecting them." The president replied that the United States position only concerned itself with tests that produced seismic readings of 4.75 or more, which Eisenhower said would come from bombs somewhere around the 20-kiloton range. Anything below that would require a separate plan.[78]

After May 1, 1960, the possibility that any agreement might be reached on a test cessation treaty diminished greatly, at least in the minds of the participating parties.[79] On that day, Soviet forces shot down a top secret US U-2 spy plane, piloted by Francis Gary Powers. Since 1956, U-2 missions had gathered information, snapped photographs, and provided the United States important intelligence on the Soviet Union that it otherwise never would have had. At first, the US position on the downed plane followed a predetermined cover story that the aircraft was a low performance, high-altitude weather research plane. Soviet Premier Khrushchev, over the course of two weeks, slowly released information to the world public that showed US statements were outright fabrications. Once the USSR revealed that it had the pilot, Francis Gary Powers, in custody, the United States could no longer deny the U-2's true purpose. Higher-ups had instructed Powers to kill himself if captured to avoid any such incidents, but in the moment, as scientific adviser to the president George Kistiakowsky put it, "he just chickened out." Kistiakowsky explained the incident simply: "the affair affected the whole tone of the administration."[80]

In the days after the Soviets shot down the U-2 plane, at first US administrators acted as if everything would be fine. On May 3, the AEC commissioners met to discuss the US position on how many inspections should be permitted in the USSR under a test cessation agreement. In short, the minutes noted that a "re-evaluation of the proper threshold for detection together with further study of the relationship between on-site inspections and the pattern and number of inspection stations within the USSR [had] presented new problems." Other issues included how and what type of monitoring system would be used (a new system installed or already existing stations used) and what would happen to the information as the stations collected it. On-site inspections could be fly-overs, follow-up ground inspections, or drilling operations. Of course, new data from the RAND Corporation indicated that increased control stations might mean that on-site inspections could be decreased, but some maintained a position that the United States should propose a high number of on-site inspections so as to have something to give up in negotiations. On-site inspections could have cost up to $6.5 million a year.[81]

By May 11, the United States seemed to be abandoning the possibility that a test cessation treaty could be accomplished during the Eisenhower administration, as the U-2 affair overrode any environmental or security concerns that might push toward an agreement. In preparation for a press conference on that day, the president received a briefing paper that counseled him how to answer questions about a recent United States announcement to resume tests. Eisenhower was supposed to respond that this announcement had been misunderstood, and that these tests were purely so that the United States could "further an improved capability to detect and identify underground nuclear explosions." Moreover, the president was instructed to explain that the tests did not represent anything new, just the ramping up of an existing program.[82] No matter the justification, nuclear tests of any sort would have made negotiating a nuclear test cessation treaty extremely difficult.

Furthermore, advocates for more nuclear tests remained in positions of power, particularly in the military. Where diplomats might have worried about public reaction to increasing atmospheric radiation, military men worried about more narrowly construed national security issues. In June, Nathan F. Twining, chair of the Joint Chiefs of Staff, sent a memo to Secretary of Defense Thomas S. Gates on the subject "Draft Treaty on the Discontinuance of Nuclear Weapons Tests." In that letter, Twining explained that the Joint Chiefs' position held "that an adequate military posture for the United States will not be attained until there is available a complete spectrum of weapons compatible with modern delivery systems which will make it possible to apply selectively adequate force against any threat." A test ban would not allow the United States to develop that spectrum of nuclear bombs. And even though the Joint Chiefs recognized that stopping tests could indefinitely maintain the nuclear arsenal advantage the United States held over the Sino-Soviet stockpile, only a treaty complete with adequate safeguards would do so. Fundamentally, the Joint Chiefs believed detonating nuclear weapons as tests played an "essential" part in maintaining nuclear deterrence and that national security should take precedent over other factors.[83]

Even without a feasible treaty in the near future, the United States continued to work toward a time when an agreement might be possible. The AEC commissioners devoted their entire July 1 meeting to discuss a proposal for a "Seismic Research Program." AEC Chairman McCone reported that after a meeting of the principals the previous day, a new proposal for seismic research existed. Policymakers grounded the program in a control method of pooling devices between the United States, UK, and USSR, with each nation sharing designs with the other two nations. In yet another instance of amalgamating geopolitical security and environmental concerns, significant issues existed with such a program, especially over whether Soviet nuclear devices might be used on US soil within the testing program and whether certain tests constituted weapons tests or not. Finally, the commissioners decided "that too much detail should be avoided in presenting the statement initially [to the Soviets], and that further details could be more advantageously determined after initial Soviet reactions."[84] Several days later the commissioners again focused an entire meeting on the seismic research program. Particularly important to those at the meeting seemed to be disassociating the Plowshare program for peaceful uses of atomic weapons from the seismic research program.[85] Yet no matter what other quandaries might have popped up from time to time, devising an effective and agreeable control system continued to dominate the US position.

A late July 1960 position paper on the issue of nuclear test detection at high altitudes provides one last example of how environmental science governed US thinking on test detection. Previous evaluations noted that a high-altitude detection system could be installed in about two years but would depend on unproven techniques and instrumentation. Moreover, any "determined violator" could evade any potential control system with existing technology. This caused the US Defense Department to recommend that the nation limit its treaty obligation only up to tests in the top of the stratosphere (30–50 km, or the detectable portion of the atmosphere) with research undertaken on higher altitudes. The State Department, however, did not agree with this position. Either way, detection techniques would have to be improved constantly to stay ahead of evaders, as US ideas of a ban still fundamentally depended on effective controls. The position paper explained that any treaty based on "unproven technique would be inconsistent with our basic principle." Eventually, the paper recommended that the Defense Department take the following positions: limit test bans to environments where they could be detected, propose a separate moratorium of no more than two years that banned outer space nuclear tests, and the Department of Defense should defer tabling the outer space ban until more negotiating could be done "concerning the coordinated seismic research program."[86] Eisenhower's briefing papers before a mid-August press conference confirmed this idea, and stated that the United States still intended to negotiate for a control system, "which would assure that no nuclear weapons tests are carried out clandestinely. However, until such a system is brought into being we have no way of being absolutely certain that the Soviet Union is not testing clandestinely."[87]

The final months of Eisenhower's term confirmed that a test ban would not be implemented during his presidency. The United States had suspended weapons tests for more than two years, since October 1958, not only to aid in negotiations but also because, in the words of the AEC, "radioactivity from nuclear tests is a source of concern in the minds of many people."[88] But mutual concern over the problems associated with radioactive fallout, by itself, could not force the two sides to come to terms. The US position still rested on issues of detection and the US worry that its own security, and by extension that of the rest of the world, might be harmed by stopping nuclear weapons tests.

In the end, President Eisenhower felt that the U-2 affair—and not disagreements about control systems—had ultimately ruined any chances the United States had to solidify any test cessation agreement, and perhaps even did more harm than that. After the incident, George Kistiakowsky met with Eisenhower, and the subject of the U-2 plane came up in their conversation. Eisenhower claimed that the scientists had let him down, but Kistiakowsky responded that scientific advisers had warned the president that the Soviets would eventually gain the capability to shoot down the planes. Instead, Kistiakowsky declared that it was the bureaucrats running the program who had let down the president. At that point:

> [The] President began to talk with much feeling about how he had concentrated his efforts the last few years on ending the cold war, how he felt that he was making big progress, and how the stupid U-2 mess had ruined all his efforts. He ended very sadly that he saw nothing worthwhile left for him to do now until the end of his presidency.[89]

But perhaps Eisenhower misjudged the situation, and a fundamental difference in positions—especially related to detection systems—meant that blaming the U-2 incident does not make sense. Instead, it is very possible that the United States and Soviet Union simply were not in the correct respective places for the agreements to be feasible.[90]

What is clear, though, is the crucial role that the natural world and scientific understandings about it played in the negotiations. The danger of radioactive fallout provided background to the beginning of the negotiation process and became a more pressing concern as these progressed, especially as poisoning the environment with further tests might damage the United States' reputation or be used by the Soviets as propaganda. The natural world, in this case, was a negotiating tool. Whether induced by heartfelt concern or political maneuvering, policymakers and negotiators should be taken at their word in this instance—protecting the environment mattered, whatever their reasons may have been. Moreover, policymakers' decisions helped enact an institutional position that valued improving environmental science as a strategy to better national security in a Cold War geopolitical context.

Beyond a desire to protect the natural world, environmental science played a crucial part in the largest point of discussion in all treaty talks—detection systems. Both sides thought testing should be curbed, at least to some significant degree, but could not agree on how such a ban might be monitored or even if it should. Representing a fundamental lack of trust in the Soviets' willingness to maintain any agreement reached, the United States insisted on only banning tests that could be effectively detected if another country violated the treaty. For tests to be monitored and detected, scientific understandings of the environment proved central. Not only did tests conducted aboveground leave telltale radiation in the atmosphere, which could be detected, but the ground itself and tremors throughout it took center stage when discussing underground tests. The whole detection argument centered on environmental knowledge and how it had to be continually improved so that detection systems could keep apace of any potential treaty violators. At their core, the Eisenhower-era test cessation talks demonstrate that the development of nuclear weapons depended heavily on how those technologies were understood within a combination of environmental and geopolitical contexts. US policymakers may have only cared about protecting the natural world for how that could protect US citizens, but the natural world itself, and scientific understandings and measurements of world systems, became hotly contested areas of conflict.

PART TWO

Nuclear Technologies by Land and by Sea

CHAPTER FOUR

Atomic Agriculture

A growing plant is a chemical factory, of course. Scientists have known this for years—but haven't known exactly what went on in that factory. They didn't know and couldn't find out how chemicals entered the plant, what the chemicals did, how they accomplished their work. So, agriculture has had to depend on trial-and-error in producing vital food.

Now agricultural science has perfected a way for studying and following plant chemicals from the time they leave the soil until they are finally deposited in the various parts of the plant. By mixing small quantities of radioactive isotopes with the soil, the scientist, with his Geiger counter, can now follow the movement of important chemicals through the whole cycle of plant life. . . . Food production, therefore, is passing from trial-and-error to certainty.[1]

—*Learn How Dagwood Splits the Atom!* (1949)

The loveable cartoon character Dagwood Bumstead is best known for eating impossibly large sandwiches, napping, and of course his beautiful wife, Blondie. During the Cold War, however, the United States enlisted the good patriot Dagwood to help teach the nation about nuclear science. In *Learn How Dagwood Splits the Atom!* (fig. 4.1), the magician Mandrake shrank our animated protagonist and his family to the size of molecules, and in their diminutive states the Bumsteads learned about the composition of atoms and how nuclear chain reactions work. The booklet not only sent Dagwood on his miniaturized journey, but also, unsurprisingly, acted as a booster for the nuclear industry.

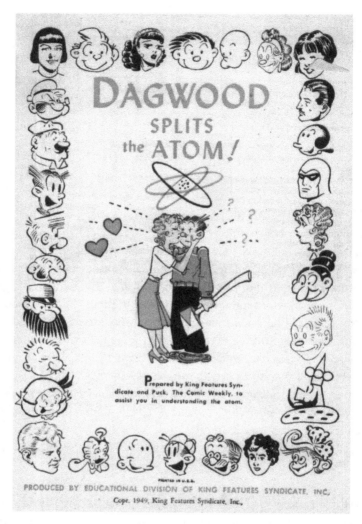

Fig. 4.1. Title Page of *Learn How Dagwood Splits the Atom!* This booklet attempted to explain, in very plain English, exactly how many nuclear processes happened and certainly is representative of nuclear boosterism. Though produced by King Features Syndicate, Inc., a comic and newspaper print syndication, author Joe Musial wrote the work with the scientific advice of Lt. General Leslie R. Groves, Dr. John R. Dunning, and Dr. Louis M. Heil. In the introduction, Groves, described as having "headed the great organization which developed the atomic bomb" (the Manhattan Project), wrote to readers, "To those who will read it carefully, this pamphlet will bring a clearer understanding of atomic energy. Many will understand what has formerly confused them. Mere words need not frighten them in the future—words such as fission, isotope, proton, chain reaction and atom bomb. This book will reassure the fearful that the future can be made bright." Joe Musial, *Learn How Dagwood Splits the Atom!* (King Features Syndicate, Inc., 1949), introduction. ©BLONDIE ©1949 by King Features Syndicate, Inc. World rights reserved. Used with permission.

As part of this mission as a booster, *Learn How Dagwood Splits the Atom!* promoted the benefits of harnessing the atom to improve agriculture. Completely outside the tiny Dagwood story arc, several pages at the end of the comic were single-page snapshots of how atomic energy had benefited, and would continue to benefit in the future, medical science, industry, and agriculture. Radioactive tracers had revealed secrets long hidden from scientists, who could now track "plant chemicals from the time they leave the soil until they are finally deposited in the various parts of the plant." A miracle technology, radioactive tracers had begun transforming food production from "trial-and-error to certainty." An image meant to portray a bucolic utopian vision reinforced the text (fig. 4.2). In the midst of a verdant landscape with perfectly delineated square fields and well-maintained buildings, a single cow stands apart from her bovine brethren, cheerily looking directly at the reader. The scene, ostensibly the end result of atomic agriculture, is as idyllic as can be imagined, reinforcing a nostalgic vision of American farming while simultaneously draping that nostalgia with a techno-modern veneer.

The implication was clear: if researchers could only understand the exact biological processes that govern how plants grow and produce food, those scientists would be able to help farmers feed the nation in a failsafe fashion. As an Atomic Energy Commission (AEC) report to Congress in the same year of the Dagwood cartoon's publication claimed, "The story of the Garden of Eden and the myth of Promethean fire find uncanny parallels in the huge responsibilities of the Atomic Energy Commission to control the unprecedented forces of atomic energy for the welfare of man."[2] With atomic energy, US policymakers hoped to turn the country's agricultural lands into a modern-day Garden of Eden, albeit with less devastating apples.

The Dagwood cartoon is emblematic of how early nuclear policymakers paid careful consideration to the ways growing nuclear scientific understandings might be applied outside of improving war-making capabilities. General Leslie Groves, former head of the Manhattan Project, believed *Learn How Dagwood Splits the Atom* would "reassure the fearful that the future can be made bright."[3] The fact that Groves saw a need for reassurance underscores the reality that a significant portion of Americans realized that atomic energy could pose incredible dangers to both human health and the environment. It is one of the era's great ironies, as historian Joel Hagen has explained, that scientists could leverage this fear. Ecologists in particular benefited, to the degree that Hagen argued, "Professional ecologists effectively used concerns over atomic energy as a convincing justification for ecosystem studies." Nuclear energy

AGRICULTURE

A GROWING plant is a chemical factory, of course. Scientists have known this for years—but haven't known exactly what went on in that factory. They didn't know and couldn't find out how chemicals entered the plant, what the chemicals did, how they accomplished their work. So, agriculture has had to depend on trial-and-error in producing vital food.

Now agricultural science has perfected a way for studying and following plant chemicals from the time they leave the soil until they are finally deposited in the various parts of the plant. By mixing small quantities of radioactive isotopes with the soil, the scientist, with his Geiger counter, can now follow the movement of important chemicals through the whole cycle of plant life.

Potash, needed by growing plants, is stored in the soil—but nobody has known how. Now science is learning the answer by following with a Geiger counter the movement of radioactive potassium atoms in the potash.

The growth of plants is known to be regulated by plant hormones. Just how plant hormones stimulate plant growth is a question which, if answered, would mean millions more bushels of food. The action of plant hormones in producing growth is being probed by radioactive atoms and Geiger counters.

A big question that has baffled science is, how does a green leaf change the energy of sunlight into the energy of starches and sugars in the plant? The scientists call this process "photosynthesis", and with the aid of radioactive isotopes they soon may find the answer.

Most of the present study with radioactive isotopes in agriculture is concerned with the nature of plants. Later this knowledge will be applied to the treatment of plants not only for healing their diseases but also for making them more resistant to pests and hardships.

Food production, therefore, is passing from trial-and-error to certainty.

Fig. 4.2. Page on agriculture in *Learn How Dagwood Splits the Atom!* This one-page section of *Learn How Dagwood Splits the Atom!* helped explain how agriculture could expect to see big gains from atomic research. Joe Musial, *Learn How Dagwood Splits the Atom!* (King Features Syndicate, Inc., 1949), 31. ©BLONDIE ©1949 by King Features Syndicate, Inc. World rights reserved. Used with permission.

became "a kind of double-edged sword" that could wreak havoc on the natural world, but also unlock "many of nature's secrets for human benefit."[4] The nation's fields and agricultural laboratories hence became sites of practical application of nuclear science where exchanges of energy, nutrients, and other material flows could be studied and put to nonmilitary use.

Nuclear research aimed at improving the nation's agricultural practices offers a revealing window into the ways in which executive decision makers during the Truman and Eisenhower administrations understood the relationship between agriculture and the atom and how they explained their understandings to other policymakers and the public.[5] While nuclear science proved important to agriculture during the 1940s and 1950s, this chapter argues that agriculture was also important for nuclear development because it provided a clearly peaceful output for atomic research. After 1945, high-level decision makers recognized they would need to justify research into atomic energy and downplay the possible harms inherent in it. Atomic agriculture not only well fit modernizing and technologizing trends in agriculture that stretched back decades, but it also repurposed nuclear science into a life-creating rather than destructive force. Likely because nuclear weapons drew so much public ire, executive decision makers intended atomic agriculture to bolster optimism that technology and greater control of nuclear energy could create a better nation. Splitting the atom in this context was a gift to the world, not a harbinger of the apocalypse as depicted on the cover of the August 4, 1947, issue of *Time* magazine where AEC Commissioner David Lilienthal's face was in front of a fiery red horse with the headline, "Is there any way out of the Atomic wilderness?"[6]

At the same time policymakers wove atomic energy into the fabric of US agriculture, agriculture became engrained in US nuclear technologies and the AEC. Agriculture has long been considered a hallmark of civilization, and the AEC intended to improve society by improving agriculture.[7] Atomic agricultural research could and did take many forms, especially including the byproducts of nuclear energy like radioactive isotopes and even radioactivity itself. What tied together these disparate research methods and atomic energy products is that each fell under the umbrella of the AEC and therefore had to be fitted into an AEC worldview that valued the expansion of nuclear science. Agriculture was merely one vehicle for that to occur.

Before going further, it should be made clear that agricultural history and environmental history have a great many linkages. In what became the opening salvo in a field-defining 1990 roundtable, historian Donald Worster's essay "Transformations of the Earth" specifically focused on "the analysis of modes of production as ecological phenomena, particularly as they are articulated in agriculture."[8] Years later, historian Sterling Evans argued for the centrality of environmental history to understanding food and agricultural history over the *longue durée*.[9] And historian Paul Sutter's 2013 state-of-the-field essay

"The World with Us" posited that agroenvironmental history was one of the "vital areas" of environmental history scholarship.[10] Historian Peter Coclanis has argued for the reverse of this as well.[11] Taking agriculture to be an extension of the natural world and worth studying through an environmental perspective not only aligns with the interrelationships between ecosystem and agricultural studies happening at the time, it also has roots in the origins of the field of environmental history itself. Moreover, nuclear technologies came along at a crucial time in the advancement and modernization of agriculture, making the time period particularly worth reflection.

Understanding this interplay between nuclear science and agriculture requires some additional background on the modernization of agriculture and the Green Revolution.[12] Farmers in the United States had begun their journey to modernized farm production in the nineteenth century, but the pace of modernization quickened after the First World War.[13] As farmers increasingly adopted machines powered by hydrocarbon fuels like gasoline, mechanical aids let farmers do their jobs easier, quicker, and with less human labor. The federal government also aided farmers with the Morrill Act of 1862 (later expanded in 1890) that created land grant colleges and eventually their attendant state extension services. After the 1920 census announced for the first time that more Americans lived in cities than in the countryside, increasing agricultural production took on a cultural imperative. With this federal commitment and ever-increasing machines came a rising industrial logic, as the transformation also had an ideological component.[14] The addition of atomic science into agriculture can therefore be interpreted as one component in a much longer history of modernization and industrialization.

After World War II, agriculture took off in an even more spectacular way, undergoing a production revolution impelled by machines, chemicals, and new plant and animal breeds.[15] Continued use of machines combined with a budding US chemical industry as tractors went hand in hand with fertilizers, herbicides, and pesticides.[16] Improved cereal grain crops like highly productive dwarf wheat and rice strains contributed to shifts in agriculture that witnessed the production of previously unconceivable amounts of food.[17] To be sure, the process did not happen seamlessly, however, and farmers made many individual decisions along the way as the process advanced in starts and fits.[18] But in the end, farming became more streamlined, rationalized, and dependent on federal support like many other business endeavors in the United States, leading to the current state of trucking cheaply produced agricultural products across the country to feed a nation that eats more and at less expense than any before it in history.[19] To wit, in 1901 the average US family spent 42.5 percent of its annual expenditures on food, while in the early twenty-first century only 13.1 percent of expenditures were on food.[20]

As the United States reached the point of over-abundant food production, a series of decisions to share the methods to such agricultural productivity coalesced into what is commonly called the Green Revolution. Previous world hunger, such as the El Niño-exacerbated fin-de-siècle drought famines that killed upward of sixty million people in India, China, Brazil, and several other countries, had elicited little attention from the United States and especially the US government.[21] With the advent of the Cold War, however, feeding the world and eradicating hunger became an important political goal.

Nick Cullather described how the Green Revolution has typically been conceived as "the greatest success in the history of foreign aid since the Marshall Plan" but instead should be thought of as "the birth of a new type of international politics." Many US leaders believed that a multitude of hungry peasants in Third World countries, particularly Asia, presented a threat to foment communist uprisings or politically destabilize the region. Feeding these hungry mouths would not only head off those problems but also help integrate Asia's population into the world economy.[22] While many place the beginning of the Green Revolution in the 1960s, the ideological tenets developed decades earlier. For example, in addition to that new philosophical approach to foreign aid, US Public Law 480 (also called Food for Peace) was enacted during the Eisenhower administration with the general goal of using agricultural surpluses in the United States to feed poor people across the world and open up new markets for US agricultural products.[23]

It did not quite work out that way. Though intended as a foreign aid solution that would put a hungry Third World into the United States' camp, the Green Revolution did nothing to change existing social imbalances. And a host of unexpected outcomes, like pesticides damaging both the environment and human health, meant that even its successes came with distinct failures.[24] In short, the Green Revolution was no perfect solution and perhaps what US planners considered to be the problem (lack of food) was more a symptom of uneven development than the problem itself. Debates over the relative merits of the Green Revolution, however, were still in the future when policymakers began serious discussion of atomic agriculture. Its promise, tied as it was to geopolitics, was foremost in their minds, and atomic agriculture proved a vital tool (and investment) for advancing agricultural production as a means to fight communism around the world.

Agricultural research with atomic energy began most prominently with the use of radioisotope tracer atoms.[25] A June 1946 press release by President Truman boldly declared, "The first peacetime applications of the results of wartime atomic research becomes immediately possible with announcement today of availability of radioisotopes for biological and medical research."[26] Even though less than a year had passed since the August 1945 bombings of Hiroshima and Nagasaki, Truman's administration already had begun to promote atomic energy as a peaceful entity. Produced from the "atom pile," radioisotopes offered scientists the ability to use "tagged" atoms—radioactive versions of common elements—to track how these atoms moved through biological processes, ecosystems, or anything else through which elements moved. Applying the tracers to agriculture seemed quite logical and, as President Truman expressed, would revolutionize biological research. The results from radioisotope research caused policymakers to champion the atom as a true boon to agriculture.

David Lilienthal, chair of the AEC, continued the president's tone of progress in his September 1947 speech titled "Atomic Energy Is *Your* Business," sponsored by the Civic Organizations of Crawfordsville, Indiana. Previously the head of the Tennessee Valley Authority, it is no surprise that Lilienthal promoted a collectivist stance when it came to the benefits of atomic energy. As the speech's title indicated, Lilienthal's first goal on that Monday evening was to contradict the notion that atomic energy was none of the US public's business, as Lilienthal considered such a stance "plain nonsense, and dangerous nonsense, dangerous to cherished American institutions and for that reason dangerous to national security." He also stressed that atomic energy and science did not change the "fundamental principles of democracy" and made the point that *"atomic energy and atomic bombs are not synonymous."*[27] Lilienthal also contended that the United States' new nuclear capabilities did not change the nation and should be made to conform to US ideals and identity, not the other way around. In this way, atomic energy's relationship to agriculture should be conceived as a supporting force and not as anything that changed farming's character.

Most of all, Lilienthal's speech emphasized the benign possibilities for nuclear energy. He reminded his listeners that the sun is an "atomic energy factory," and continued, "Atomic energy is not just another new gadget, nor just a new weapon, however powerful and devastating. We are dealing with forces as fundamental to your life as the force of the sun, the forces of gravity, the forces of magnetism." By comparing splitting the atom to the processes of the sun, Lilienthal morphed harnessing atomic forces from something danger-ous and arcane into something natural and common that his audience dealt with every day.[28] Nuclear energy was not new, dangerous, and full of peril but instead similar to a warm sunbeam on your face or any other natural force that humans could bend to their will for benefit.[29]

Even if splitting the atom was not new or dangerous, Lilienthal still cham-pioned atomic energy for its ability to throw "a clear light upon some of the oldest mysteries of life." For example, humans could now answer the ques-tion of "how does a stalk of growing corn use the rays of the sun to manu-facture its products into energy-giving food substances?" Atomic energy in agriculture therefore shaped national destiny and buttressed the principles on which patriots had built the United States.[30] In addition to producing re-search that could serve the dual purposes of both agricultural and national improvement, Lilienthal and the AEC had the benefit of nuclear research that did not have to be associated with atomic weapons in any way (even if some research did go into using contaminated livestock, crops, and agricultural land as a form of bioweaponry).[31]

In addition, popular notions about technology's rightful place in agricul-ture undergirded Lilienthal's contention that nuclear energy could play an im-portant role in agriculture. From John Deere's metal plow in the middle of the nineteenth century to Henry Ford's automobiles, trucks, and tractors, ag-riculture received a significant technological boost in the century before Lil-ienthal's speech that caused productivity to soar and prices to drop.[32] Hence when Lilienthal's audience received the good word about the atom, they likely readily accepted the idea that technology fit into agriculture like sunshine and good weather. If the atom could provide the newest piece of equipment for farmers, it should be welcomed with open arms.

In an address to the Annual Meeting of the American Farm Bureau Fed-eration later that same year, Lilienthal connected atomic research to agricul-ture even more explicitly. He explained, as his number one point, "No one in this country has a greater stake in the vigorous development of atomic en-ergy, and the consequent increase in knowledge of the fundamental laws of Nature, than you who day after day work most closely with nature—the farm-ers of America." Since he thought farmers had such a high stake in atomic development, Lilienthal's second point followed closely when he contended that farmers needed to stay informed of atomic energy discoveries and peace-ful uses of the atom.[33]

With the speech, Lilienthal, and by extension the AEC, underscored his desire to distance atomic energy from atomic bombs in the minds of his audience members. He attacked the "myth" that atomic energy could only be used as a bomb or as a weapon, arguing instead that "*Nothing could be farther from the truth*" because peaceful nuclear research had opened a "new world of knowledge." Indeed, treating atomic energy and atomic weapons as linked would cause the United States "to fall into an even deeper pit of error. We will grow forgetful of the *true sources of America's strength*." For Lilienthal, in implicit contrast to the Soviets, the US got its strength not just from its military prowess, "but rather in the spirit of this nation, in *the faiths we cherish*." The United States was set apart by "a deep sense of stewardship to our Creator, the Father of us all; and when that is no longer strong within us we are weak and we are lost, however heavily armed with weapons—even with atomic weapons—we may be."[34] Invoking stewardship added additional stress to his point that atomic forces were "fundamental," comparable to gravity and magnetism and could be harnessed as such. Anything that perpetuated the myth conflating the atom with military might would, in his opinion, only work to deprive "you and yours of the peaceful fruits of this discovery," implicitly subverting the will of God.[35]

Not until page six of his speech did Lilienthal get back to his main point "that the farmer and the farm family have a very special stake in the wise and vigorous development of the science of the nucleus of the atom, for peaceful purposes." He even compared the incredible stores of atomic energy to farm energy, saying, "the energies that produce great poems, that build churches and homes, the energies from which spring such noble ideas as our Constitution and Bill of Rights. That energy has been stored up in the plants of the field, and in the tissues of the animals that feed on your pastures; thence it comes to men." Farms had produced food from the atomic energy of the sun for millennia, and farmers represented "the trustee and steward of that never-ending miracle by which the atomic energy of the sun becomes chemical energy and then human energy."[36] With this reasoning, farmers held an important stake in the development of atomic energy and its application in peaceful endeavors. Farmers made possible all the United States' great history and ideas by nourishing the bodies that produced these marvels, and the country needed them to help continue this great legacy. Hence the AEC needed farmers, the trustees and stewards of the sun's atomic forces, to help support its atomic energy research agenda.

To Lilienthal, the difference between "a modern American farm and a backward poverty-stricken farm" was knowledge, and "in this country the farmer has seen that the scientist is his partner, his companion and friend." Lilienthal's message held a clear implication—if providence (or the AEC) gave farmers, the "custodian of the sun's energy and the forces of growth," the opportunity to do something like develop nuclear power they surely would. The AEC chair gave the example of phosphorous to help explain why the wise farmer would want atomic science developed, just as they had previous technologies that improved farm production. He elaborated that, even though it cost a great amount, US scientists could produce radioactive phosphorous. By using tagged radio-phosphorous, scientists could help "in a way never before possible chart the changes that occur in matter in the process of plant life and growth. In your behalf, the researcher can gain new and important knowledge of how plants convert the sun's energy into life energy on this planet." Clearly this represented the farmer's *"big stake"* in nuclear development.[37] Since scientists worked on the behalf of farmers, per Lilienthal's own words, it seemed only logical that farmers would support their efforts, as supporting scientists truly was, in effect, supporting themselves.

Near the end of his speech, Lilienthal brushed aside any concerns his audience might have had over exactly what the uses of the atom in agriculture might be, encouraging that the breakthroughs would be significant. He reminded them that many prominent scientists, like Gregor Mendel, had been unsure of what their research might mean when they began, though that research eventually proved fundamental to farmers. Lilienthal noted that harnessing the atom might also improve agriculture through pest control, pointing to an upcoming conference on the subject at Alabama Polytech at Auburn (today Auburn University). And while radiation might not be useful directly as fertilizer or in foods (though research would continue on this subject), agricultural improvement remained "one of the glorious promises of atomic science. It well may help to solve one of the most vexing problems of humanity—how to keep food production in pace with the growth of the world's population." Lilienthal ended his speech by claiming, "Trained as are no other group of men in the discipline of understanding and working with and through natural forces, endowed by the very nature of your calling with both persistence and patience, you American farmers are uniquely qualified to play a leading part in realizing the beneficial possibilities of this new force."[38] Thus farmers, using atomic agriculture, would play a pivotal part in US foreign aid plans in the future as the United States reconceived of its world role as helping poor, underdeveloped, and hungry countries become modern, fed, and prosperous nations.[39]

Lilienthal did not stand alone in his ideas about the importance of farmers, and indeed federal focus mirrored such notions. Historian Jenny Barker-Devine noted that campaigns by many federal agencies used rhetoric that "placed farmers and rural residents on the front lines of the Cold War," in contrast to these people's typical moral and geographical distance from the urban centers most likely to experience nuclear attack. US administrators at the time described farmers as essential to the Cold War effort, both for the agricultural products they produced and the moral bedrock they provided. And as the danger of radioactive fallout caused fear that rural areas too might be subject to danger (from that fallout), distributing scientific information to rural communities became even more important.[40] Hence farmers played a crucial role in the development of the atom both rhetorically and in terms of direct research.

The lofty claims in Lilienthal's speeches served as harbingers of the AEC emphasis on agriculture. When the magician Mandrake shrunk the loveable Dagwood and his family, the purpose was to assuage fears about nuclear science and educate the public about the possibilities of the atom to improve the nation. Lilienthal attempted something similar in that he wanted to calm a fearful public and create great hope for the results new technology might bring. Beyond bucolic ideals, the AEC also counted on agricultural development via atomic energy to help distance some atomic research from atomic weapons—as Lilienthal worked so hard to say in his two speeches, the two were not inseparable. The commission also demonstrated a focus on agriculture in its reports to Congress, and the January 1949 report showed the commission studying both how living creatures absorb radiation and also using radioactive tracer atoms to follow life processes.

Even though David Lilienthal warned farmers in 1947 that radiation would not be useful as fertilizer, during the 1948 growing season the AEC-supported experiments in fourteen states on nineteen different crops to see if radiation could be successfully yoked to boost plant growth. In November 1948, Dr. Alexander from the Department of Agriculture, which had received AEC funding for its research, reported to the AEC commissioners on "the results of a study of the effects of certain radioactive fertilizers on plant growth," summarizing there was "no indication of any increase or decrease in plant growth" due to radioactive fertilizers.[41] Even with unsuccessful experiments in using radiation to induce plant growth, the AEC planned more experiments for 1949. The January 1949 report of the AEC questioned, "Does Radiation Stimulate Plant Growth?" After calling the idea "nearly as old as man's knowledge of radioactivity," the report emphasized that experiments using radioactivity as a fertilizer were "quite separate and distinct" from fertilizer tracer research on "the rate and volume of movement of various fertilizer materials in the soil, their absorption into the plants, and their accumulation in plant parts." The commission expected those tracer studies would "solve practical problems of fertilizer application which are of direct dollars-and-cents interest to farmers, fertilizer producers, and farm machinery manufacturers."[42] Even if radiation did not work as a fertilizer itself, research using radioactive isotopes could make existing fertilizers work better and unequivocally save farmers, and through them the rest of the nation, money.

The July 1949 report further portrayed improving agriculture as one of the commission's goals. In a section on "Radiation and Life," the commission described all of the ways that humans had learned about radiation, peaceful and violent, helpful and harmful. The report explained, "Radiation attacks, disrupts, and destroys the delicate electro-chemical balance in the atoms, molecules, and protein combinations within the bodies of living things. As a result, it damages and kills the cells of which atoms and molecules are a part. If enough cells are destroyed, the whole organism—plant, animal, man—is severely injured or dies." In spite of this statement, though, the AEC continued its program on radioactive fertilizers.

It is unclear where the logical disconnect occurred. Clearly knowledge existed that radiation harmed living things, but somehow this fact did not manifest itself in the cognizance that radiation might not be successful as a fertilizer. The partition of knowledge between researchers persisted either way. Researchers tested the same crops as in 1948 and still found no beneficial effects. The AEC also ran agricultural experiments on studying cattle exposed to radioactive fallout dust, understanding how fertilizers feed into plants, and many other smaller programs such as studying photosynthesis, mineral nutrition, and improving fungicides and herbicides.[43]

To help meet its goal of improving agriculture, in the latter half of 1949 the AEC expanded its programs studying the effects of "atomic energy and its products" on plants and animals, both in AEC installations and at the various colleges and universities through which it contracted. Not all of this research centered on agriculture. For example, at the Hanford plant, sited on the Columbia River in Washington, researchers determined that some organisms, such as plankton, could tolerate radiation much better than could other animals, such as humans. In the field of agriculture in particular, research continued on how plants take in and utilize fertilizers, a program that the planners hoped could "mean a saving of thousands of dollars to the farmers of the country." Specifically, researchers determined that crops use certain forms of phosphate better than others and that soil acidity plays an important role in how certain crops process that nutrient.[44] This point about saving money would come up consistently in AEC reports. Research was and is an expensive endeavor, and therefore any program that could claim to pay for itself in savings passed directly onto the taxpayers would be much easier to justify.

Despite increasing knowledge of radiation's dangers, research into radioactive fertilizers continued, demonstrating a deep commitment to atomic agriculture and a desire to prove that radiation itself could be beneficial to life forms.[45] In 1950 the commission reported that too much radiation could slow tomato plant growth. Its studies found that if tomatoes received 20,000 roentgens total at a rate of 150 an hour, the plants would suffer ill effects.[46] The next AEC report clarified the seemingly commonsense (even then) position that radiation would hurt plants: "Experiments gave no indication that radiation could improve growth rate or yield, but in large doses caused marked damage to both."[47] Fortunately for taxpayers, not all AEC-supported research proved so fruitless.

Use of radioisotope tracers continued to form a crucial component of the commission's research and helped the AEC show how atomic technologies and radiation could be beneficial. For example, research delved into how cattle interacted with their environment, particularly how the ruminants broke down feed and converted that to milk. Other investigations used radioiodine to study plant growth regulators and also looked into how mealybugs affect pineapple plants, using radioisotopes to study salivary secretions of the pests. Research even tested radioactive weed killers to determine how plants interacted with the chemicals. Further studies used radioisotopes to look at how plants absorb nutrients into their roots, transport them throughout the plants themselves, and then deposit those nutrients in the various plant structures.[48] Radioisotope research proved diverse and robust, and the AEC continued its research programs in 1951. That year agriculture and animal husbandry research advanced especially on the subjects of the metabolism of cows, fertilizers, and plant nutrition.[49]

The January 1952 AEC report to Congress contained the largest section yet on the atom and agriculture, with dozens of pages under the heading, "Atomic Energy and Its Applications in Plant Science." Important for understanding the commission's motivations, the report claimed two broad objectives in supporting research in plant science, one related to radiation safety and the other to improve agriculture as an industry. The first encompassed determining "the effects of radiation and radioactive products upon plants in order to broaden scientific understanding and to aid manufacturers and users of atomic energy in adopting measures to safeguard life and property." In short, the AEC wanted to help protect "crops and other property" from the damages radiation might present, as research "is necessary to cope with circumstances that may follow atomic explosions." Focused on protecting the United States during an atomic bomb attack, knowing how plants and animals reacted to radiation exposure might be vital to the nation's long-term survival. The second reason for studying the atom and agriculture was to "help in the application of atomic energy products and techniques to fundamental and applied research with plants" for the benefit of the United States' people and industries. During the fiscal year that ended June 30, 1951, the AEC budgeted $20.6 million on the entire field of biology and medicine, with $1.3 million (around $13 million in 2019 dollars[50]) researching plant life.[51] While not the commission's primary focus, clearly the expenditure represented a significant emphasis on the matter.

The first research listed in that January 1952 AEC report focused on "Intense Radiation and Plant Development" and provided an endpoint to previous research. Different from past investigations, the AEC did not present this inquiry as any sort of fertilizer program. Instead, pertaining to objective one of the plant science research program, the research focused only on how radiation affected plant growth so that the AEC would know how plants might be affected after a nuclear blast. In general, the experiment produced mixed results. For example, on tested potatoes, some grew malformed, but others failed to sprout at all. Interestingly, these latter potatoes did not rot in the ground—irradiating the potatoes seemed to preserve them. This information would be important in the future. Fungi tended to better handle radiation than plants, so using radiation as a fungus control seemed impractical—dosing the undesired fungus with enough radiation to kill it would do more harm to the plants to be protected than to the attacking fungus.[52]

Finally clearing up previous investigations into radiation being used as a fertilizer, the report stated, "Claims that radioactive fertilizers would increase crop yields have been discredited by repeated tests. Even small amounts of radioactive material used for 'tracer' research in plant studies may—unless care is taken—damage the plants and cause error in observed results." The most obvious question is thus, "why did research into radioactive fertilizers continue for so long?" The answer to such a query is made even more vexing considering the United States Department of Agriculture (USDA) claimed back in 1914 that radioactive fertilizers did not work. Even with such a judgment, further tests continued with USDA backing until 1944. Though the USDA had discredited radioactive fertilizers decades before, agricultural scientists considered the question anew after the Hiroshima and Nagasaki bombings because observers claimed in the aftermath there had been "greatly increased crop yields" near the cities. Indeed, even John Hersey, in his harrowing *New Yorker* articles (which ultimately became the book *Hiroshima*), noted that weeks after the atomic bomb many flowers bloomed, seemingly from the radiation. "It actually seemed," he wrote, "as if a load of sickle-senna seed had been dropped along with the bomb."[53] In hindsight, however, it became clear that something else caused those bountiful harvests. Repeated findings showed that if radiation had any effects on plant growth, those effects would be negative, either killing the plant or stopping it from growing (or never growing in the first place).[54]

At this point, the notion of radiation as a fertilizer seemed officially dead, but the fact that it held sway for as long as it did in research programs is important for what it says about the AEC. Clearly research and practical experiences had demonstrated the great dangers of radiation. Though Hersey had reported about flowers in bloom in Hiroshima, even the US public knew that radiation represented a real threat to human health after *Hiroshima* described in vivid detail the devastation wrought by the first atomic blast in Japan.[55] But if decision makers could show that radiation had benefits or could even be healthy for some organisms in certain contexts, the moral position of creating radiation—a position the AEC, as the bomb creators, knew was extremely tenuous after the public's interest in Hersey's *Hiroshima*—would change dramatically. A desire to undercut some of the moral dilemmas attendant to nuclear science might go some way toward explaining why AEC research plans held onto the idea that radiation might function as a fertilizer so long after it was known that that radiation proved harmful to plants. Thus forays into the potential benefits of radiation did not end with the acknowledgment of its dangers, as the AEC still worked to show how radiation could indeed be a positive force. For example, the January 1952 report to Congress also stated, "Although fertilizers that depended on radioactivity for their action proved useless [notice not harmful, but merely "useless"], radioactive tracers are showing how conventional fertilizers can be used more efficiently and economically."[56]

In contrast to failed studies into radioactive fertilizers, the AEC chronicled radiation's harmful effects quite clearly, which reinforced the commission's need to find peaceful and helpful aspects of atomic energy. Beyond its obvious effects on living tissues, radiation also seemed either to kill soil microorganisms or make these much less effective. Many microorganisms around plant roots help fix nitrogen, an element vital to plant growth, and thus if radiation killed these microorganisms it could be especially damaging to plants. Moreover, the January 1952 report to Congress recognized strontium 90 (Sr^{90}), an isotope produced as fallout from nuclear explosions, as "potentially the most biologically hazardous of the fission products."[57] In contrast to Sr^{90}, reactor cooling water seemed safe despite its radioactivity, because most of that is either short lived or diluted, even if some might get absorbed by plants or algae. Effects on plants also depended on exposure, with those effects especially noticed in the growing sections of plants.[58]

Research into how fallout radiation might harm US agriculture also reflected a deep understanding of new geopolitical realities. In August 1949, the Soviet Union detonated its first nuclear weapon, which meant that the United States suddenly had to contend with another country potentially unleashing an atomic blast upon it. With this new reality came a desire to know exactly how the nation might be affected. Experimentation into what might happen to US agriculture after nuclear attack—discerning how fallout and other radiation affected plants—became an even more important part of the US atomic energy program.

Studies on genetics and radiation also spoke to the AEC's mission to better understand living beings through radiation. For example, inquiries found that corn exposed to less than five röentgens of radiation exhibited no appreciable effects, but exposure between 5–55 roentgen caused mutations proportional to the radiation dose. Mutations can occur naturally, and, though some are beneficial, an overwhelming majority end up being negative (at least from the perspective of the individual organism). If radiation in a controlled laboratory setting could speed up the rate at which mutations occurred, beneficial mutations could be created, discovered, and isolated much more quickly than if humans left natural processes to their own devices. Radioisotopes also helped make possible research into tree and crop diseases, insecticides, herbicides, and photosynthesis.[59]

As the Eisenhower presidency began, no great changes in agricultural research occurred from the AEC perspective, though a focus on peaceful uses of the atom increased. As the files from Eisenhower's 1952 presidential campaign show, developing atomic energy into a true industry formed an important part of Eisenhower's platform. Citing the need both to "improve the atomic arsenal" and continue "to probe the frontier of knowledge," soon-to-be President Eisenhower claimed during his 1952 campaign that nuclear energy should be viewed as past leaders had considered the steam and internal combustion engines. "Both of these opened vast new field[s] in the development and application of energy and were considered by some to be so dangerous that their need should be carefully and rigidly controlled by government." Eisenhower cautioned against being afraid of advancing nuclear technology and instead contended that present decision makers needed to be as prescient as their predecessors and support the development of this new technology, atomic energy, and all its beneficial advances. In this mindset, properly developing atomic energy certainly would create great developments in many fields, including agriculture.[60]

These campaign speeches made sense in the context of an Eisenhower administration that tried to base agriculture on free market ideals to a greater extent than had his Democratic predecessors. Eisenhower's selection of Ezra Taft Benson for the Secretary of Agriculture reflected this shift in emphasis toward free market ideology. A well-known conservative, Benson believed that agricultural problems of the 1950s stemmed from overproduction by farmers in previous decades. Benson's policies, especially cutting holdover price floors from the 1930s, combined with other modernizing impulses in US agriculture and led to over half of the country's 5.8 million farms failing. Edward and Frederick Schapsmeier claimed this happened due to "business failure, particularly among the small, inefficient operators."[61] In hindsight, it is clear farm failures occurred as part of a trend toward larger industrial farms and away from family farming (something Eisenhower, the rural Abilene, Kansas, native, likely would have supported in rhetoric as part of a Jeffersonian ideal, but obviously not in practice). Decisions during the Eisenhower era represented notions that agriculture should be considered a business, and atomic energy improving agricultural technology well fit a mantra later popularized by Richard Nixon's Secretary of Agriculture Earl Butz, who famously quipped, "Get big or get out."

With the nuclear industry and the threat of nuclear war in mind, AEC-sponsored investigations continued into how plants dealt with radiation. Those experiments studied how plants grew in soil containing concentrations of "fission products" (such as fallout like Sr^{90}) equal to the maximum fallout observed at nuclear blast sites. Growing radishes, barley, oats, cowpeas, and ryegrass, researchers found that strontium was indeed the radioactive element most likely to be absorbed by plants, but this occurred at a lower rate in soils rich with calcium (remember strontium and calcium function very similarly in biological processes). When cattle ate plants that contained radioactive fallout, cattle absorbed 25–30 percent of ingested strontium, with about 25 percent of that reaching the bone. Researchers said this bone contamination would only be a hazard to humans if they ingested the bone splinters that might get into meat, though.[62] Other experiments measured how radiation sickness affected animals and used radioisotopes as tracers to study how tropical crops absorbed potassium.[63]

A 1954 speech by Richard Bradfield, head of the Cornell University Department of Agronomy, emphasized the already present and growing importance of agriculture to atomic policymaking. Though not an executive branch policymaker himself, Bradfield's speech nonetheless well laid out the current state of research and policy at the time. Speaking at a New York meeting sponsored by the Atomic Industrial Forum, Inc., Bradfield stated clearly, "In spite of our unprecedented increase in population, our agriculture has been able to keep up with our continuously expanding needs. It is now easily possible for 10 percent of our population to produce all the food and fibre which our entire population needs. Probably never before in the history of the world have so few people fed so many so well."[64] Where just a few decades earlier during the Great Depression policymakers considered overproduction the problem, by the mid-1950s that overproduction meant prosperity at home through cheaper food prices and the ability to save lives and bolster world opinion of the United States abroad by feeding impoverished peoples around the world.

While reminding his audience that agricultural surpluses had been rare for most of human history, Bradfield contended that a steady food supply is important to both national security and peace. Agriculture thus transitioned from merely being the way we feed ourselves into being the way that the United States could help support geopolitical stability and perhaps even set up itself as the leader of that new world order. Atomic agriculture could do this because radioactive tracers allowed a scientist "to follow the meanderings of his atoms" and ensure productive agriculture, defined by Bradfield as the combined product of good soil, varieties of crops being suited to environment (better seeds), and the reduction of the threat of plant diseases and insects (all of which could be improved by radioactive tracers).[65]

Continuing this emphasis on productive agriculture, Bradfield reported that at least 30 percent of the recent increase in US agricultural productivity had been from fertilizers, but continued, "We know that some of these fertilizers are not being utilized effectively." Here he singled out phosphorous use in fertilizers because he thought the country needed "wise use" of its phosphate reserves. USDA and Oak Ridge laboratories tested fertilizer intake by plants with radioactive phosphorous tracers, which enabled scientists to tell how much phosphorous came to a plant from fertilizer and how much came from the soil's natural phosphorous. The experiments also enabled plant physiologists to understand better the role of phosphorous in a plant's internal functions.[66]

Bradfield also talked about the importance of previous work on radiation and mutations so that his audience could further appreciate the gifts atomic energy provided agriculture. He explained that mutations are frequently useful to plant breeders, and "seem to be produced under natural conditions by radiations which reach the earth from outer space, the so-called cosmic rays." But, as has been explained earlier, these do not occur very often and breeders frequently wish they could speed up these mutations—radioactive isotopes help speed up this process. The Gamma Field, located on Long Island near Brookhaven, represented the best example of this. There, radioactive cobalt got lowered into the ground by remote control when needed, and then researchers planted crops in concentric circles around the cobalt. Researchers studied the resulting crops, and "occasionally" one of the resulting mutations from exposure to the radioactive cobalt proved beneficial. Bradfield stated that already one promising crop, "a mutant of oats," had been produced that "seems to have resistance to one of the most destructive diseases which attack this important crop."[67] Disease-resistant mutant oats were proof that nuclear science had much to offer agriculture.

Bradfield ended his speech by trying to temper enthusiasm for atomic agriculture because radioactive isotopes did not represent "a complete panacea for all the agricultural ills of the world." And yet, they did have the power to help scientists "unravel many of the mysteries which have so far eluded them. It will enable them to trace these elements from fertilizer to the soil, to the plant, and through the plant to the animal and then to man."[68] The biggest takeaway message for Bradfield, however, seemed to be that improving agriculture with the atom meant more than enhancing food production and continuing to produce fantastic surpluses—it meant a policy decision about the security of the nation. In this way, nuclear science, agricultural science, and national security were all linked together.

By and large, AEC research reflected a position that atomic energy benefited agriculture and by extension the nation, furthering its own research agenda into nuclear science. For example, the AEC reported in July 1954 on studies of how radiation affected plant growth and reproduction. It claimed an objective of testing "the feasibility of producing useful mutations by means of ionizing radiations in plants, shrubs, and trees normally propagated asexually." These experiments produced several varieties of disease-resistant plants. The AEC continued earlier experiments on irradiating potatoes as well, with the intention of determining what it takes to prevent these from spoiling.[69] Another project studied calcium and magnesium content on twelve Wisconsin and Illinois farms in relation to how atoms got exchanged.[70]

Strong research clearly demonstrated nuclear science's benefit to agriculture, but public promotion of atomic agriculture would be needed for agriculture to benefit nuclear science. In 1954, James Hagerty, White House Secretary, communicated with organizers from the Toledo Council on World Affairs. Those organizers wanted to bring to Toledo, Ohio, the atomic energy exhibit from Oak Ridge, Tennessee, and arrange for sixty thousand schoolchildren to attend it free of charge. One of the nine key points of that exhibit was "Atomic energy in the processing of feed and as it affects plants and domestic animals."[71] Teaching children about atomic agriculture, in contrast to several speeches to professional organizations, seems to have had little purpose other than public promotion.

Relatedly, AEC support of agricultural studies using radioisotopes continued to occupy a prominent role in justifying how those programs produced interrelated benefits for both atomic energy and agriculture. Experimenters paid particular attention to studying "the intake of radioactive materials" by livestock, including tissue distribution, absorption, retention, and excretion.[72] Intake by plants also received study. The AEC reported to Congress that "Knowledge of the effects of fission products on plant growth and reproduction is important in evaluating health and safety aspects of atomic tests and production operations of nuclear reactors." Such findings proved that agriculture could help better understand the effects of atomic bombs, emphasizing that agricultural research helped more than just the production of food and fiber. Research also confirmed that fallout products tended to act like other elements—strontium like calcium, cesium like sodium or potassium, and so forth. Other research, at North Carolina State College, used radioactive tracers to show that corn obtained about 70 percent of its nutrition from the top ten inches of soil and peanuts about 87 percent from the top ten inches. How nutrients get absorbed depended on the specific plant and its root distribution pattern.[73]

Radioisotopes also could be used for much more than uncovering how plants absorbed nutrients, which helped push atomic agriculture into new realms. The tracers made possible inquiries into how a rubber plant produces its valuable product, and then enabled tracking that produced rubber to see how it broke down and degraded. In addition to tracers, experimentation continued into how plants absorbed fallout products. The AEC also reported that the ratio of calcium (and thus strontium) plants absorbed seemed identical to the concentration of other chemicals (ammonium acetate in particular) in the soil in which the plants grew. Researchers continued to use radiation to produce mutations in plants, hopefully improving crops' physical characteristics and disease susceptibility. And if immunity could not be created to diseases or pests, then radiation might be used to control those pests. Research showed that nematode worms might be controlled by radiation, but of course too much radiation would prove injurious to the plants on which the worms feasted.[74]

The combination of technological optimism and boosterism of atomic energy in relation to agriculture continued in 1956 with the "Report of the Panel on the Impact of the Peaceful Uses of Atomic Energy to the Joint Committee on Atomic Energy." While technically a report in the legislative branch, Robert McKinney, former assistant secretary of the Interior under Truman, chaired the panel. Additionally, Carl Hinshaw (R-CA) from the Joint Committee on Atomic Energy (JCAE) had previously sent Eisenhower a copy of "The Contribution of Atomic Energy to Agriculture and Medicine," which further demonstrates a close linkage between the legislative and executive branches in this instance. President Eisenhower had responded that he was "fascinated, as is everyone, by the potentialities of this great new era into which we are so rapidly moving."[75]

The panel's report devoted chapter 5 entirely to agriculture and argued, "Peaceful uses of atomic energy in the field of agriculture are a significant addition to the many other modern methods of improving farm technology." Not only did atomic agriculture mean "increased productivity and lower costs for individual farmers," but the report argued that improved agriculture also gave the United States a "dramatic opportunity to lead underdeveloped, undernourished nations to higher living standards."[76] Only by sharing food production techniques with impoverished nations, by cultivating a burgeoning

Green Revolution, could US planners safeguard the Third World from communist influence and keep those nations secure from destabilizing influences. Atomic agriculture could therefore play a significant role in defining the United States' place in the world.[77]

The panel expected the United States to accrue many benefits from atomic agriculture, and they wanted to promote that to the nation. The power of the atom could help scientists learn more about life processes of the plants and animals in agriculture, how best to use fertilizers, insecticides, and medicines, and create new plant varietals better adapted to their environments, more resistant to diseases, and "tailored to mechanized cultivation and harvesting." In short, using the power of the atom would "add great impetus to the technological revolution in agriculture." Readers were to "expect higher farm output, more flexibility as to the crops and animals produced, and ultimately more varied diets at lower costs."[78] Language like that made harnessing the atom in agriculture not only a foolproof way to improve the nation's resources—a way one would have to be foolish not to support—but, if placed in the context of the Green Revolution, also became a moral imperative for helping to improve the world.

Plant breeding provided a dramatic expression of how radiation could be beneficial to living beings just as the AEC had hoped earlier experiments into radiation fertilizers would.[79] Scientists could do this by using atomic energy "to speed the evolution process." This implied that radiation mutations were not unnatural, but instead merely helping natural forces work a little faster than these might on their own. Exposing plants, animals, or insects to radiation made it possible to create new varietals more quickly and replace natural selection with human choices. As Lloyd Berkner once quipped, "It is as though, for evolutionary purposes, we had collapsed a thousand years into one."[80] The report further explained that only a small percentage of the new "variations" would be good, and scientists still had to winnow these from the unhelpful ones so they could be "put to work on the farm." The report closed the section by boldly claiming, "At least on a laboratory scale, the day of the tailormade plant seems close at hand."[81] Already on hand, though, was the day of using hopes for the atom in agriculture to push research agendas.

Atomic boosterism caused the AEC to cast even seemingly negative experimental results, such as the development of new blights with increased virulence, in a positive light. The report claimed that the creation of these new blights under controlled conditions allowed geneticists to breed plants resistant to the new pestilence, preparing the plants and farmers for these new blights before they appeared in the field.[82] Such a statement assumes one of two things—either the blight created under laboratory conditions would at some point get out into the larger world and plague crops that way, or natural selection and evolution are sure to produce the same or a similar disease on their own. Assuming the first (laboratory release) would not happen, any assumption that the second would occur represents an understanding of evolution that is far too linear and progressive to be accurate (just because laboratory conditions produced one blight does not mean that natural conditions ever would have).

Other parts of the panel's report seem like science fiction. The report claimed that researchers could duplicate many of the steps involved in photosynthesis, meaning that a time was "within the realm of possibility" in which humans would not depend on plants "to produce edible energy in the form of starches, sugars, fats and proteins," but this could instead by done chemically on a commercial scale. And if other boosterish claims were not so farfetched, they still presumed a great deal. For several pages the report made claims about how atomic energy would help produce more food on fewer acres at a lower cost. Since a "principal fact of the American way of life is that it is based on abundance," creating even more abundance with food would only enhance the lives of the nation's citizenry, as surely low food prices would stay low (how such production might hurt farmers went unmentioned).[83] Whether fact or reality, because policymakers believed that excessive production meant consumer prosperity, more food would lead to ever-lower prices on the shelves and improved lives of the nation's citizenry.

This report also explicitly insisted this new knowledge and technology could help the United States feed the world, emphasizing a perception that the United States' role in the geopolitical realm had changed. It stated bluntly that the United States "can help the undernourished peoples of the world have more to eat" so long as more research, education, and work occurred, as there would be "no miracles" without these. The report finished with three recommendations: the United States needed to keep researching; those dealing with the farm surplus problem should take into account that atomic developments will exacerbate the problem; and an exploration of the humanitarian benefits that could result should begin immediately. The third point held particular importance, as "only in this way can the United States bring to bear atomic contributions to agriculture, so as to demonstrate our historic sense of international humanitarian leadership."[84] This "historic sense" of humanitarianism held the concurrent purpose to help establish the United States as a world leader in contrast to the Soviet Union. If the United States could help feed the world it would have significant leverage in the Cold War court of world public opinion. Following the chain of connections, since US nuclear science created atomic agriculture and atomic agriculture could help feed the world and improve the lives of world citizens, policymakers figured US nuclear research could be considered a benevolent force for improving the world. By tying atomic agriculture to a US global imperative, policymakers crafted space for the world public to view US nuclear science as a life-giving endeavor.

The AEC continued to use agriculture as a public demonstration of the benefits provided by atomic energy. In its January 21, 1956, issue, *Science News Letter* ran an article titled, "Atoms Vital to Agriculture." That article cited Dr. Williard F. Libby, AEC commissioner, as claiming that the US economy may get "as big a boost from the use of atomic energy in agriculture as it will from atom-generated electricity." Though he did not provide any sort of timetable for when gains could be expected, Libby's "low" estimate was a $210 million-per-year benefit. In general, Libby propounded atomic benefits to agriculture in fertilizer studies, pest control, and preservation.[85] The article thus effectively served as an atomic booster and gave a broader audience to Libby's voice, and by extension the AEC.

In July 1956, the commission claimed that the radioisotopes used for research and production in industry and agriculture already repaid the United States "a dividend of several hundred millions a year" on monies invested. Of course, it added, that such focus on money ignored "the value to mankind of these substances as scientific tools, diagnosis, treatment and scientific study of human diseases and their consequent alleviation of human misery."[86] Strangely, at the same time the AEC made such incredible claims as to the value of the atom in agriculture, its reports on agricultural research waned a bit. The January 1957 AEC report to Congress only reported on the use of radiation to inhibit photosynthesis as 100,000 roentgens of gamma radiation temporarily reduced that process in wheat to 25 percent of normal.[87] It is unclear how such research might have proved compatible with previously stated goals.

It did seem clear to AEC officials, however, that radioisotopes were of incredible use to the US atomic energy program and the public needed to know that. A February 1957 AEC report prepared in advance of hearings by the Congressional Joint Committee on Atomic Energy echoed Libby's previous estimate that the savings in agriculture by radioisotopes might reach $210 million per year. It remains somewhat unclear as to how the AEC arrived at those

numbers, however, even if it claimed "no reason to doubt the order or magnitude of these figures." As an example of this, the material estimated that
radioisotopes saved the nation around $500,000 to $1 million per year in fertilizer studies, "based on spot checks." Moreover, the report emphasized that
the knowledge gained in studies with radioisotopes needed to be passed onto
state, federal, and country agricultural organizations, as ultimately farmers
would need the findings for these to be of any real use. In short, "there is every reason to anticipate that when this translation can be accomplished and
made available to the nation's farmers, the estimated potential savings of $210
million per year can in a large measure be realized within the foreseeable
future. Such savings could be reflected in an improved farm productivity at
lower unit costs not only in this country but also in other nations."[88] The report clearly implied that if only the knowledge gained by scientists with the
help of radioisotopes could be put in the hands of farmers, the nation would
become even more prosperous.

At this point, much of the research intended to eventually help farmers, at
least as conveyed to the public by the AEC, was lists of the previously known
ways that atomic energy could improve agriculture. Reiterating the importance of radioisotopes and their use, the July 1957 report to Congress listed
as the major benefits to agriculture, "(a) better placement and application of
fertilizer, (b) new and improved growth regulators, herbicides, etc., (c) improved measures against plant diseases and fungi, (d) better knowledge of
animal nutritional needs, (e) improved measures against animal diseases, (f)
better insect control through sterilization, insecticides, and information on
migration and hibernation, and (g) new or improved varieties of plants and
breeds of animals." In fact, these benefits had proved so valuable that agricultural use, in conjunction with use by medicine and industry, had created
such demand for radioisotopes that supply could not keep up with demand.[89]

One new avenue of research pursued by the AEC in the late 1950s centered on irradiating seeds and crops to produce beneficial effects and continued the theme of searching for positive benefits of radiation. Just as earlier research had accidentally discovered with potatoes, irradiating, if done at proper
levels, could significantly improve the shelf life of agricultural products. Too
much irradiation, though, could be harmful to seeds. Some research found
that after seeds had been irradiated, stored, and then planted, radiation damage could be three times higher than if scientists only irradiated and planted
them. Water, oxygen, and heat exposure before and after irradiation also affected how seeds performed.[90]

This is not to say that previous avenues of research did not continue as part of the AEC's plan to improve the nation through atomic agriculture. Emphasis on radioisotopes and the amount they saved the nation continued, with special attention paid to the gains made in "broadened knowledge and improved management" of both crops and livestock, including a greater control over the diseases and pests that afflicted both. Better fertilizer use and improved insecticides and herbicides also derived from research, with "benefits in sight from widening experiment with plants and animals."[91] The January 1958 report later elaborated that researchers made these gains with "essentially a byproduct of atomic energy activities—the radioactive isotopes of the natural elements created in nuclear reactors." Radioisotopes also helped scientists create soil moisture and density gauges, useful in both agricultural and industrial processes. [92]

Even when not explicitly focused on improving agriculture, AEC research frequently found grounding in it. Other projects focused on "the impact of various atomic energy activities on *man's environment*." The AEC intended these studies to better understand "the balance" between all life forms, whether they live in land or water habitats. The report claimed that the answers gained would assist decisions about the extent to which agriculture and other atomic energy activities "may occupy an area and lead to general benefit rather than detriment."[93] More direct inquiry into agriculture continued as well.

Significant research in livestock and insecticides persisted, especially in using radioisotopes to track biological processes in the studied creatures. Such research continued to provide a public display of the atom's gifts to agriculture. For example, using radioisotopes, scientists uncovered that some fatty acids absorbed by cattle in their digestion are used to form milk sugar lactose, while others are used principally to make butter fat. Researchers uncovered other technical information about digestion as well. In general, studies tagged parts of a cow's feed and then traced those bits to see how cows transformed feed into milk. For insecticides, by using radioactive tracers, scientists determined not only exactly how pesticides affect pests but also how much toxic residue made it into and onto raw agricultural commodities. Experiments also successfully led to the eradication of the screwworm fly in controlled tests on the island of Curaçao in the West Indies. Since screwworm flies caused damage to Southeast livestock of around $10 million a year, finding a way to combat the insects seemed important. Radioisotopes also provided insight into how herbicides affected plant growth regulators, helping scientists study the herbicides 2,4-D and 2,4,5-T.[94] However, the most heavily pushed research occurring at the time related not to using radioisotopes but instead to using radiation to change the composition of foodstuffs.

By the end of the decade, irradiating foods and seeds at precise levels occupied much of the ink received by atomic agriculture. Irradiating foods represented a different process than using radiation to create beneficial mutations, which the AEC reported were "being found in sufficiently high numbers to justify continuing efforts."[95] In contrast, irradiating foods could extend shelf lives of previously perishable products. In February 1960, the commissioners of the AEC met and discussed the establishment of a radiation processed food program. The Interdepartmental Committee on Radiation Preservation had proposed a conservative investigation into the potential of irradiated foods building upon a similar Army study from 1953. At that time, the Army performed experiments on twenty-six types of food, particularly focused on unrefrigerated preservation for up to a year. It found that only certain meats—beef, pork, poultry, and ham—fit the desired specifications.[96] (Army plans even advanced far enough that in 1956 the Army Quartermaster Corps built a reactor just for food irradiation.[97]) Thus while atomic agriculture could improve the lives of the nation's citizenry by increasing the nation's food stores and serve the national security mission by feeding a hungry world, it also could enable the US military to conduct even longer troop deployments than previously. If "an army marches on its stomach," then having food that would last for a year without refrigeration might keep soldiers marching for a long time if they were cut off from supply routes.

Though the Army program certainly found some success, there had been no testing on civilian foods as such would have been out of the military's purview. The commissioners, however, decided that the AEC should support the Interdepartmental Committee's program so that civilian food could be tested. More than seeking to fill a hole in a research program, the AEC thought the food irradiation program fit the AEC's mission (along with the Atoms for Peace program) of finding peaceful applications of atomic energy. Logically, the AEC should then pursue the research because of its "unique knowledge and competence" concerning the involved technology. The Army had experienced storage and logistical issues with their irradiated foods, especially related to bacterial contamination of foods irradiated at high levels and then stored for extended periods. Canning had been necessary to solve this problem, but discussions did not seem to find this a particular problem for future AEC experiments. Eventually John McCone, the AEC chair, declared that the program "held promise for revolutionary developments for the food industries of the world." The commission then approved $115,000 in their budget for research in fiscal year 1960, with $500,000 planned for the fiscal year 1961 budget.[98]

About a month after that meeting of the AEC commissioners, the AEC made its plans public when the Research and Development Subcommittee of the JCAE held a hearing on a food irradiation program. At that hearing, Richard Morse, director of the Army Research and Development program, presented the Army's revised research program on preserving food through irradiation. This program had seemed sensible and been well received but focused on high-level radiation sterilization and preserving food for a year. In contrast, the AEC's civilian program would emphasize low-dose "pasteurization" to extend the shelf life of perishable foods—civilians did not necessarily need meat that could sit in their pantry for a year at a time, but having fruits and vegetables stay fresh longer before spoiling would have been nice. The one snag in the commissioners' plan seemed to be that low-dose radiation might not be commercially available for five to ten more years. No matter, the Joint Committee wanted to push the programs "because preservation of food by radiation was a dramatic program easily understood by the public." The commissioners agreed, and their only concern was how the program might appear to a public that had been promised rapid results—results that might be hard to deliver so quickly.[99] The AEC did not stand alone in a desire to show the world the benefits of irradiation.

After the AEC decided to support irradiation research, scientists conducting the research also had an interest in seeing those programs succeed. Because of this, Dr. C. J. Spears of Oak Ridge Atom Industries, Inc., asked President Eisenhower to take some of his company's irradiated flower and vegetable seeds to plant on the president's farm. Eisenhower was a bit of a farmer himself, having grown up in rural Kansas. He once mirthfully remarked, however, "You know, farming looks mighty easy when your plow is a pencil, and you're a thousand miles from the corn field."[100] Hopefully the irradiated seeds would make Eisenhower's life a little easier. Spears' representative explained that the president planting the seeds himself would "awaken the people of the US further to the many benefits that have been afforded them as the result of the efforts of the Republican Party in promoting the peaceful uses of atomic energy."[101] As could be said for the program of using atomic energy to improve agriculture more broadly, irradiating food and seeds meant more to its proponents than merely a way to better handle the nation's food production.

By the end of the 1950s, agriculture had put down deep roots as an important part of the nation's atomic energy program. John McCone's letter of resignation to Eisenhower just a few weeks before the end of Eisenhower's presidency helps show this fact. McCone's resignation included a statement titled "Eight Years of Progress in Atomic Energy" and deemed the advancement of the nation's nuclear programs "substantial." In that statement, McCone listed among his successes radioisotope progress in fertilizers and weed killers, radiation in plant genetic improvement and pest control, and generally improved agriculture.[102] Looking back at the end of his term as chair of the AEC, McCone counted atomic agriculture as one of the accomplishments of his tenure.

After 1960, significant research into the applications of atomic energy in agriculture continued, particularly by the United Nations' Food and Agriculture Organization (FAO) and International Atomic Energy Agency (IAEA). In many ways modeled after the United States' AEC, the IAEA developed after Eisenhower's 1953 "Atoms for Peace" speech and in 1964 even teamed up with the FAO to create a special FAO/IAEA Joint Division. Historian Jacob Darwin Hamblin chronicled this tale and showed a confluence of modernizing principles, science, technology, international politics, and agriculture. In his estimation, the IAEA "succeeded in reshaping the UN toward a particular technological path of modernity," often at the expense of the FAO and the scant resources of developing countries, all the while brushing aside any significant critiques of its activities. As Hamblin described, the IAEA's *"raison d'être* [was] to promote a particular set of technologies"—promoting peaceful uses for nuclear technology—and not necessarily foster agricultural development. "To abandon food and agriculture," Hamblin argued, "would have been to undermine a crucial component of 'Atoms for Peace' that specifically targeted the developing world." Thus a story that began with research sponsored by the AEC in the mid-1940s continued history after Eisenhower left office.[103]

In the end, using nuclear science to improve agriculture showed several things about the United States. First and most obvious, it functioned as a way to improve the nation's agriculture and agricultural production, even though by the 1950s one of the most serious problems the nation's agriculturalists faced was how to deal with the incredible surpluses of food they already created. Yet those in power repurposed overproduction as a way for the United States to feed a world that policymakers conceived of as being filled with hungry people in need of US aid (for both their own good and that of the United States). Particularly with radioisotope tracers that helped unlock many biological mysteries, US agriculture harnessed the atom quite success-fully. But using nuclear science in agriculture had another goal than nobly ensuring that food production passed "from trial-and-error to certainty" as the Dagwood cartoon claimed.

Perhaps even more important than its obvious purpose of improving farm-ing, atomic agriculture functioned as an important way to show how splitting the atom could do more than unleash death and destruction. By emphasiz-ing the nonviolent applications of nuclear energy, programs that attempted to improve agriculture allowed the Atomic Energy Commission and the execu-tive branch to say to the public, with good reason, that they desired peaceful applications of atomic energy. Clearly the first worldwide application of split-ting the atom had been horrific—no matter your side during World War II, the bombings of Hiroshima and Nagasaki terrified almost everyone to some degree. But through agriculture, something fundamental to modern human existence, policymakers hoped to refocus nuclear science from its more sin-ister applications. Showing that using atomic energy could be peaceful dra-matically changed the AEC's purpose and transformed the organization from death dealer to life bringer. In this way, research into agriculture using atomic energy could be just as useful to the AEC as it was to fields and farms.

Studying atomic agriculture also opens a window into the perceived place of agriculture in both the United States and the world at the time. Agricul-tural modernization with mechanization and chemicals found its logical next step in atomic agriculture, as the atom represented the newest technology that could be put to work for the good of farming. This let US farmers pro-duce food more cheaply and efficiently, meaning US citizens got more bang for their buck in grocery stores, all while supporting the continued rise of agribusiness. Internationally, anxieties about feeding the world (necessitat-ing increased food supplies) also meant that the United States could manu-facture a new place for itself in the world—a role not only as world food sup-plier but also as a distributor of knowledge proverbially teaching the world to fish rather than fishing for it. Both of these facets of food production—at home and abroad—aided atomic agriculture in bolstering nuclear technolo-gies and furthering their development, which created a sort of feedback loop between the atom and agriculture. Supporting atomic research thus meant furthering agricultural modernization and a nascent Green Revolution, and frequently the inverse of that held true as well. In this way, atomic agricul-ture helped integrate agricultural environments into both national and inter-national societal structures—the Green Revolution not only became a part of how US society functioned, but even more it affected governments on a worldwide scale.

It should be noted, however, that the US nuclear program likely did cause significant harm to the nation's agricultural production. Economist Keith Meyers has attempted to measure the degree to which radioactive fallout affected fields and farm animals, finding that wheat, corn, sheep, and cattle experienced statistically significant losses. He estimates that atmospheric nuclear tests at the Nevada Testing Site from 1951–1970 probably caused a loss of about 236 million bushels of winter wheat and two billion bushels of corn. Sheep decreased by 2.6 million head and cattle by 2.3 million head. The total cost of those losses was nearly three billion in contemporary dollars or nearly thirty billion in 2016 dollars. Showing the wide reach of fallout from nuclear tests, Iowa, Kansas, and Nebraska—and not any states bordering Nevada—suffered the greatest losses.[104]

While bomb improvement and production may not have stopped at any time during either presidency, research into agriculture allowed the United States to take a morally superior position. Not everyone believed in the idyllic ends that such programs might achieve—helping the United States feed the world. But the ostensible ends proved less important than the fact that it allowed the United States to advance its nuclear program under peaceful pretenses and in doing so brought very real benefits to researchers and farmers. Fundamentally, atomic agriculture held dual purposes—agricultural improvement and the advancement of an argument that nuclear energy should be considered a benign entity and not a harbinger of death. This doubly purposed research means we need to revise not only our understanding of what atomic research meant for the environment during the Cold War but also recognize that many organizations might be willing to improve the environment if it also means improving their own public image.

CHAPTER FIVE

From Affluence to Effluence

In late 1947, Harvard University President J. B. Conant spoke about "The Atomic Age" at the National War College. Although his speech mostly focused on military matters, near the end Conant delivered somewhat of a throwaway line that showed exactly what many decision makers thought about the fate of radioactive materials. Ruminating on the world eventually ridding itself of atomic weapons, Conant explained, "In the last stage all existing stocks of plutonium and U-235 [fissionable uranium] would be dumped into the sea or 'denatured' so that the material would not be available for atom bombs."[1] Hypothetically, this sounded like a fine idea. As long as fissionable materials had been "denatured" so that the ores could not be used in atomic weapons, these posed no military threat to humans. The second option, depositing all nuclear materials in the ocean, meant humans hypothetically avoided dangerous radiation. Such a sentiment, however, demonstrated ignorance of the political, social, and especially ecological realities of nuclear waste.

At one time, humans saw the environment as "sublime, powerful, eternal, and inexhaustible." However, as Hamblin has chronicled in his studies of nuclear dumping into oceans, the natural world "became in the twentieth century a fragile entity apparently drained of is resources and life—a vulnerable earth greatly in need of protection or control."[2] In broader terms, when most humans previously thought about waste disposal they believed, as the old adage goes, that "dilution is the solution to pollution" and dumped copious trash of all sorts into the seas because they conceptualized an all-powerful natural world that could not be harmed—the planet, especially its oceans, was simply too big ever to become truly polluted.[3]

After the rise of environmentalism, however, humans began to view the environment in much more cautious ways, and a new status quo considered the natural world as a delicate balance that needed to be protected. As Hamblin described, "In the 1950s, leading oceanographers viewed the ocean as a sewer, using language that might have led to the professional ostracism of an aspiring marine scientist just a couple of decades later."[4] Thoughts about nuclear waste disposal during the Truman and Eisenhower presidencies thus reflect a position that would seem nonsensical in the political climates of later administrations—policymakers found the environment useful, important, and worth understanding, but did so in the context of seeking how best to fill various land and seascapes with as much waste as possible without affecting the bodies of US citizens. Nuclear technologies by their very nature produced a great amount of radioactive wastes, and it is impossible to separate the use of technologies from the wastes such uses produced.[5] Examining nuclear waste disposal is crucial to understanding nuclear technologies, because we cannot take an ecological approach to nuclear technologies without examining how the end results of many nuclear processes ended up with radioactive waste. Dealing with that waste required executive policymakers to utilize environmental science, and therefore it is worth studying the role those scientific understandings played in their related policymaking.

Nuclear waste disposal proved to be one of the most consistently troubling outcomes of many atomic processes. Because unmitigated nuclear pollution held great potential to harm US citizens, nuclear waste disposal forced US policymakers to support research in environmental science and include it in their deliberations. Doing otherwise would have endangered the nation. When examining the potential nuclear waste disposal site at Yucca Mountain, public policy scholar Allison Macfarlane argued for the coproduction of scientific knowledge and policy; the two occurred at the same time and each influenced the other to a degree that, in the end, science and politics became indistinguishable.[6] What Macfarlane chronicled on a small scale also happened on a grander one. Policymakers eventually recognized that if they were going to have a nuclear program, they would need to handle the waste that came from it, and safe, effective nuclear waste disposal was simply impossible without embracing environmental science.

Even with significant scientific resources available, nuclear waste disposal remained a significant challenge for US decision makers for several reasons. First, each type of nuclear waste, different in both form and harmfulness, required a different sort of handling. Finding an appropriate disposal solution therefore meant finding many appropriate disposal solutions. Second, and perhaps more important, policymakers had to combat their own intellectual biases about the planet while simultaneously balancing human and environmental health against the very real need to deal with the radioactive byproducts produced by the nuclear technologies deemed necessary and vital to the nation. In doing so, policymakers, especially at the Atomic Energy Commission (AEC), treated radioactive waste just like they would any other trash, except with the added dimension of radiation. To them, nuclear waste disposal started from a default position of dumping the wastes where other rubbish might go and then attempting to solve the problem of radiation. In essence, policymakers found it difficult to shift their way of thinking from older paradigms and thus attempted to shoehorn nuclear waste policies into already existing modes of thought. Despite access to scientific research about the interaction between radioactivity and the environment, and even possessing the necessary tools and power to change the conceptual models they might utilize for waste disposal, executive decision makers during the time period never proved capable of moving beyond an "out of sight, out of mind" attitude.[7]

During the early Truman presidency, policymakers were less concerned with solving problems related to disposal of used nuclear products and more interested in distributing nuclear products for governmental and commercial use.[8] Perhaps the most powerful demonstration of that notion is the archival holdings in the Harry S. Truman Presidential Library, where the offhand quip in J. B. Conant's speech was the only mention of nuclear waste disposal found by this author. Jacob Darwin Hamblin has even explained that, during the early Korean War, Congressman Albert Gore, Sr. (D-TN) caustically advised President Truman to solve the country's nuclear waste problems by using the wastes "to 'dehumanize' a belt across the Korean peninsula. The dangerous wastes from plutonium processing could be put to good use, [Gore] said, and the president could avoid the political repercussions of using an atomic bomb."[9] Fortunately for Korean and world environments, the technology to do so was not entirely feasible.

A step-by-step chart of the production process for nuclear technologies from the January 1949 Semiannual Report of the AEC offers a telling indication of AEC priorities and mindsets.[10] The chart appears to be fairly comprehensive at first glance, but in actuality it elides almost as much as it explains. For example, the process of mining is essentially ignored, and instead it is assumed that the raw ores almost magically arrive to the processing plant via boat or factory. More important, nuclear waste disposal is not mentioned at all. While top policymakers were aware that nuclear waste needed to be handled in a safe and effective manner, they paid little attention to it through the 1940s.

Indeed, the full July 1949 Semiannual Report of the AEC to Congress contained little more than a page on nuclear waste disposal, with that page focused heavily on how radioactive waste might affect human bodies. The report claimed, "In setting the [safety level for humans], the problem was less that human drinking water might become contaminated than that people might eat animals that drank water in discharge streams or fish that fed on micro-organisms that had absorbed radioactive material."[11] More information on decontaminated radioactive water was needed.[12] The report's statement thus fell in line with early understandings of radioactive fallout that emphasized human bodies. Polluted streams, fish, microorganisms, meat animals, and oceans only became a problem if humans might possibly ingest some of the radiation that had entered those biological and ecological systems.[13]

By the summer of 1950, however, the AEC had embraced at least a slightly increased emphasis on radioactive wastes. Its July 1950 report to Congress, for instance, included a section on "Environmental Safeguards" that offered a perfunctory recognition that the government needed to develop solutions to the problems attending nuclear waste. "The Commission endeavors to safeguard areas surrounding atomic energy installations," it claimed, "under the same mandate that directs it to protect workers in the program." The report therefore promised the AEC would set "Permissible levels of radiation released from routine operations into the environment [extremely low]—at or below the levels of background radiation under many natural conditions."[14] Even so, the fear of environmental contamination remained firmly focused on possible threats to human bodies. And no matter the precautions, AEC actions frequently produced radiation and contaminated various products with that radiation, necessitating that the commission do something to ensure the safety of humans and the environment.

In 1950, the AEC considered only two methods of controlling radioactive wastes to be viable options (not producing the wastes in the first place was not one of the options). The first possibility involved concentrating radioactive products so that these could be stored in select places where humans might be least affected (for example, cast into concrete and then sunk deep into the ocean). The alternative involved "mixing the material with so much nonradioactive material (air, water, or a stable isotope of the same material) that it [would] be harmlessly dispersed."[15] In the latter scenario, radioactive effluence got treated just like many other pollutants—radioactivity could be put directly into the air or water so long as a sufficient enough supply of the diluting agent existed. To the degree that this second solution implied that as long as the AEC diluted radioactive wastes enough these presented no discernible harm to humans, it evinced a mindset that saw radioactivity as being no different than other hazards

Paired with these general strategies, the AEC also carefully measured the environs surrounding its production facilities to ensure that disposal plans safely worked. Sites like Oak Ridge in Tennessee and Argonne Laboratories in Chicago needed frequent monitoring so that the AEC could be sure it mediated any dangers. Of course, once dumped into the environment, radioactive waste had to be guarded to keep out anyone who might go near it, and the local environments near the dumping site also had to be monitored. One of the best ways to prevent dangers, then, involved carefully choosing AEC sites to minimize the chances of any incident. When choosing a reactor location, "The AEC determined that such reactors should be tested on a large reservation of public land—preferably of submarginal value for farming and ranching and not suitable for future agricultural, mining, or other development— whose very extent would serve to guard the population of the surrounding area against potential hazard. The geology of the site was of importance; the earthquake risk had to be small."[16]

Yet AEC policies supported almost contradictory conclusions, as the commission cared deeply about environmental contamination but did not necessarily care if that contamination harmed the environment. That is to say, the AEC showed little early concern that its radioactive waste might harm the plants or animals in and around dumping locations. The commission did, however, care that such radioactivity might eventually make its way through natural systems into human bodies. For example, the AEC studied each site carefully to account for unique characteristics of each landscape—radioactive products surely would behave differently in a desert setting with underground water than at the Hanford plant on the Columbia River in Washington State. And yet, for all that careful monitoring, the most important measurements concerned how humans might be affected. For example, the subsection on "*River Studies*" held that for humans to be affected by excess radiation in fish, a person would have to eat a hundred pounds of these fish in one sitting, or ten pounds a day for a very long time "to get any appreciable dose of radioactivity." The fact that the fish themselves were tainted by radiation that might affect their own biological processes was inconsequential.[17]

Other plans offered even more startling proposals that reflected the ways in which policymakers thought about radioactive waste disposal much the same as traditional waste disposal. One AEC-supported project looked into disposing radioactive iodine and phosphorous used in medical research directly into the public sewer system. The study, conducted at New York's Mount Sinai Hospital, found "no danger to sewage disposal workers" because the sewer system diluted the radioactive products to a sufficient degree. The AEC further declared, "Plumbing fixtures through which isotope wastes had passed were dismantled, tested, and found below any degree of radioactive contamination that might be hazardous to plumbers working on the fixtures."[18]

Such a practice seemed fine on a conceptual level, and research findings also found no detectable danger, but given the context in which such proposals were offered—one in which no one truly knew what safe levels were—the ideas were brazen. Even if the scientists conducting the research were correct in their assessment of the immediate dangers, no data existed on what might happen if humans received low-level exposure over the course of a decade or two (indeed, how could such data exist considering the atomic age was less than a decade old?). The AEC thus brought into this situation, and others, a style of thinking in keeping with contemporary scientists and government officials that proved unable to move beyond such thought patterns to appraise atomic energy as something new and distinct that required eschewing previous assumptions.[19] Whether pouring radioisotope tracers into the sewers was ultimately safe or not, the AEC did not have a sufficient basis to make a judgment either way.

Continuing AEC research showed that the commission did not necessarily think that it had the problem under control, even if the organization did think its plans moved in the right direction. The July 1952 report to Congress sounded very positive and proactive, claiming, "Research, development, and investigations in sanitary engineering were advanced by AEC contractors during the first half of 1952 to obtain more efficient handling and disposal of wastes at lower cost and to secure better information on the environmental aspects of atomic energy operations." In short, AEC research programs frequently studied whether traditional methods of waste disposal could be used to deal with radioactive wastes, particularly high-volume, low-level waste. For example, at Johns Hopkins University, experiments tested whether conventional incineration could safely dispose of wastes containing radioactive phosphorous. That study found 90 percent of the radioactivity went into the ash with the remainder depositing in the stack and in small particles in the smoke.[20] Another evaluation declared that burning low-level wastes in isolated areas seemed like a good idea and was the cheapest way to dispose of products "without health hazard."[21] Burning, an imperfect waste disposal solution under even the best circumstances, represents one of the ultimate displays of the "out of sight, out of mind" mentality. Once incinerated, burned materials do not disappear but instead go into the air, soil, and water, typically transformed into different states and materials via the chemical reaction of fire. But fire would not destroy the radioactivity of those products. Adding radioactive ash to the many existing problems the AEC faced did not seem to distress researchers very much, likely because of how they conceived of fire as a disposal instead of displacement solution.

Plans still called for much low-level radioactive waste to be put directly into bodies of water, but policymakers began to fund environmental science research that would improve how that took place and better understand what happened after the dumping. At the University of Texas, researchers tried concentrating liquid radioactive wastes into algae, which would then be removed from the water by rotary vacuum filters. This would not diminish the amount of radioactivity involved but would reduce the amount of radioactive liquid to manage.[22] Other work put radioactive tracers into water to determine how long the radiation lasted in rivers. For example, in New York's Mohawk River, "preliminary analyses indicated that, under the test conditions, in roughly 5,000 feet of stream travel, the radioactivity concentration at the outfall was diluted to essentially background or harmless levels." Yet again, disposal plans drew upon the idea that "dilution is the solution to pollution." None of this research meant that the AEC felt it had the issue under control. The January 1955 report to Congress declared, "The disposal of radioactive waste is a major problem in the atomic energy program."[23] Where indifference had largely characterized nuclear waste disposal policymaking in the late 1940s, by the mid-1950s a fundamental change in conception had occurred. Not only had environmental science improved to the degree that waste disposal could not be ignored, but the ever-advancing US nuclear program simply produced a great deal more radioactive waste by 1955.

High-level radioactive wastes presented the most significant problem. Chemical plants that processed irradiated fuel elements constituted the main source of such products. The AEC considered many different types of disposal, including ocean dumping, underground holding, pumping into wells, and incineration.[24] Yet, no matter how much scientific knowledge AEC-sponsored research produced, no perfect solutions existed. In general, the AEC took three primary approaches to the problem—fix the fission products in other mediums for easier storage; selectively remove the worst parts so that the bulk could be more easily handled; or discharge the highly radioactive wastes as they were into holding tanks or specially selected geologic formations.[25] None of these options could do anything to actually diminish the amount of radioactivity contained in the effluence, even if the methods hopefully could avoid any potential damage to human health or landscapes outside of the dumping grounds.

In the latter few years of the 1950s, focus on dealing with radioactive wastes increased so much that the AEC declared, "The problem of handling and disposing of radioactive wastes runs through the entire fabric of nuclear energy operations." Reminding readers that matter in any state—gas, liquid, or solid—could emit radioactivity, the January 1957 report to Congress summed up the issue clearly when it claimed, "Because of the long life of some radioactivity, the ability of radiation to cause injury to human, plant and animal life, and its potential danger as an environmental contaminant, the safe handling and final disposal of wastes is important to the successful application of nuclear energy to peaceful uses." Thus the AEC had several objectives in dealing with the radioactive products: develop better and cheaper ways to handle and dispose of the waste; determine how much natural systems would dilute wastes and lessen the required treatment; learn more about "fundamental phenomena"; aid integration of nationwide agencies; and assist concerned state and local officials.[26] In short, though the whole program could be improved, from the nitty-gritty technical aspects to the larger, structural features, the AEC had not only recognized the challenges involved by 1957, but the organization had also recognized that the problem's inherently environmental dimensions would require increasing environmental scientific knowledge.

One thing the AEC made clear, however, was that once radioactive wastes had been disposed of out of sight, it still took a long time for these to be safely out of mind. For example, workers could bury radioactive wastes, but facilities still needed to erect fences to limit access and monitor nearby waters and soils. At the Hanford processing facility, as another example, after cooling water had been put in a storage basin to reduce some of the radioactivity and finally returned to the Columbia River to be diluted, that river needed continual study to ensure no ill effects occurred. At the Oak Ridge facility, workers excavated three pits "in the relatively impervious Conasauga shale" and between 1951–1957 dumped more than four million gallons of low-level waste into "open seepage pits," necessitating downstream monitoring to ensure drinking water safety. Since the production of every gallon of processed uranium also created between one-tenth and one gallon of high-level liquid waste, merely storing such liquids in tanks was not "a final economical answer.

On the other hand, sufficient dilution probably [was] not available in nature for any safe, continuing dispersal to the environment." Researchers thus attempted to find other solutions, such as heating the liquids to very high temperatures until these became a dry oxide powder (which could be packaged as a solid or mixed with clay and fused in a kiln to form a ceramic mass). Other possible solutions included discharging the liquid into subterranean salt beds or salt domes between five and fifteen thousand feet in depth or pumping the liquid deep into the sea where planners and scientists thought that little sea life and slow circulation would prevent damage to humans.[27] Either way, waste disposal remained an unresolved issue in 1957. Policymakers realized that their choices could have serious ramifications and thus required frequent reevaluations to be sure that no problems arose.

Even though nuclear waste constituted a serious conundrum, not all wastes were created equally, as whether the waste was liquid or solid, high or low level, could make a significant difference. In 1957 the commission declared, "The handling and disposal of solid wastes have at no time constituted a serious technical problem." To justify such a claim, the AEC reported test dumping thirty miles off of San Francisco's coast. The Scripps Institute for Oceanography studied that site and "tended to confirm that waste disposal there has produced no harmful effects." Liquid wastes, however, especially of high-level radiation, "remained the major technological problem in disposal." The best the AEC could do with that fluid was to store it in tanks, and to that date the commission had placed sixty-five million gallons containing millions of curies of radioactivity in tank storage.[28] (Even a thousandth of a curie can be fatal to humans in the right circumstances.) Thus while the AEC may have "solved" some of its radioactive waste problems, others remained significant hurdles.

The January 1958 AEC report to Congress contained a section on "Sanitary engineering research" that reviewed research primarily focused on waste handling and disposal, the water supply, and environmental sanitation. For low-level wastes, researchers tested biologic sewage treatments, and results indicated that 70–90 percent of "low-level mixed fission products can be removed." And even easier than disposing of low-level radioactive wastes itself, the AEC had licensed seven commercial firms to do disposal for the US government. Such disposal was "generally limited to handling small quantities of radioactive waste material. The wastes are disposed of at sea, are stored, or in some instances are returned to the commission for permanent burial. The hazards, both operational and long term, are comparatively slight." Such solutions would not work on "high-level residues," unsurprisingly, and the previous tripartite research into converting wastes into an inert solid, selective removal of specific isotopes, and direct discharge to selected geological formations continued. Most of that research, however, was not entirely viable, so most high-level liquid waste went directly into underground storage tanks.[29] Even with plans for either drying radioactive liquid waste to "a less hazardous, noncorrosive solid product" or possibly drilling into a salt formation at the Naval Air Station at Hutchinson, Kansas, high-level radioactivity products remained a serious problem.[30] The AEC would soon discover that high-level wastes were not its only problem.

By 1958, the commission had reached a full-fledged recognition that its environmental-science expertise was still insufficient to deal even with low-level disposal. In response, the AEC furthered its relationship with the National Academy of Sciences' Committee on Oceanography by forming a special subcommittee to examine "the feasibility of establishing a limited number of *new sea disposal sites* in the Atlantic Ocean and Gulf of Mexico for use in commercial disposal of low-level radioactive wastes. If feasible, new sites will be recommended closer to the coast-line than the presently recommended 100 miles or more offshore."[31] Doing so represented a concession that the AEC needed a proficiency in marine sciences that it simply did not possess. Partnering with an external agency provided the quickest route for acquiring that know-how.

In addition to farming out research, the AEC also continued to distribute licenses for waste disposal, and as of the end of 1958, "8 licenses were in effect, 6 for waste disposal in the Atlantic or Pacific Oceans, 1 for storage, and 1 for packaging and returning wastes to the Commission." The AEC decided not to spell out "precise details for waste disposal" in guidelines to these companies because there are so many "varied and complex technical problems" that giving leeway seemed more appropriate. The application process for ocean disposal did require a great deal of information, however, which gave the AEC at least the illusion of control even if it took a fairly laissez-faire approach after it had distributed a license.[32]

In July 1959, the AEC's assurances were not enough to placate the nation's citizenry, and a *Washington Post* article on radioactive waste contamination in rivers caused the commission concern. That article reported that the Department of Health, Education, and Welfare (HEW) sought to end radioactive contamination of US rivers and streams by uranium refineries. It claimed that about "half of the 28 ore-processing plants now in operation are dumping radium and other waste products into rivers in the West," with some levels as much as twenty-two times the maximum permissible radiation levels. Arthur S. Flemming, the HEW secretary, had scathing criticisms for the AEC and vowed that rivers needed to be both cleaned up and studied.[33]

AEC chair John McCone referenced that article in a commissioners' meeting on the day the article ran and said he was "seriously concerned about the growing volume of criticism [the] AEC was receiving on the problem of radiation contamination." Reports at that meeting claimed that the AEC had essentially been doing its due diligence by inspecting uranium milling operations and sending out notifications of noncompliance when necessary. Moreover, the commissioners noted two different factors at play in the situation. First, they claimed that river contamination does not tell the whole story, as duration of exposure mattered, and if the rivers were cleaned up soon "no harmful effects [would] result." The second point, in a bit of political maneuvering, was that while AEC responsibility covered regulating the radioactive level of effluent and dust the mills produced, "condition of the rivers as a whole is the responsibility of the Public Health Service." Other fears concerned "public misunderstandings concerning AEC policies on ocean disposal of radioactive waste." In the end, the commissioners decided that they needed "an integrated organization within the AEC to efficiently administer the entire waste disposal program and to be capable of effectively allaying the mounting public fears about this situation."[34] An event a few days later would show that the AEC did indeed have reason to fear public concern.

In mid-July 1959, the *Providence (RI) Evening Bulletin* reported on, as one angry resident described, "tentative plans for disposal of quote low intensity atomic waste unquote close to Rhode Island Coast."[35] Christopher Del Sesto, Rhode Island's governor, wrote to President Eisenhower in a fit of disquiet, "Any action of this kind would seriously affect Rhode Island's attraction as a vacation area and might also endanger the marine life for which the state is renowned." Del Sesto continued, "I feel that too little is presently known about nuclear waste to accept a proposal such as the committee on Oceanography has offered, and I therefore respectfully request that you intercede in behalf of Rhode Island" and stop the program.[36]

The AEC response to the Ocean State's Governor Del Sesto claimed that the commission had "direct responsibility for control of this activity" with no present plans for using or approving the sites without more research by a variety of groups. It ended, "Please be assured that in our consideration of these matters, protection of the public health and safety, and conservation of our natural resources will always be of paramount importance."[37] Another concerned citizen called the plans a "patent disregard for the welfare of humanity." He argued, "If the government of this country can expend billions of dollars on the development of atomic weapons and processes it can and must include in that budget funds for safeguard against a fate more horrible than most men can imagine, which can and probably will result from those weapons and processes." Underscoring the idea that the nuclear age helped birth modern environmentalism, that man concluded his message to the president by saying, "Contamination of Earth is a one way street."[38] These letters emphasize not only public worry about dumping plans, but also how such activities might affect the natural world and through it affect human health. Particularly, they highlight citizens concerned about the state of scientific knowledge and whether the AEC knew enough to follow through with its plans. The AEC, however, continued its plans for ocean dumping, which shows that public concern could only go so far and that the seas still represented one of the best places to discard nuclear wastes.

On the heels of Del Sesto's letter, a special legislative note from the AEC to the White House highlighted a hearing the following week by the Congressional Joint Committee on Atomic Energy (JCAE) on a National Academy of Sciences (NAS) report titled "Radioactive Waste Disposal into Atlantic and Gulf Coast Waters." The commission claimed that, particularly, "The JCAE is concerned over Congressional and public apprehension generated by this report and also anxious to preserve its jurisdiction in the field of atomic waste disposal."[39] That report was from the NAS's Committee on Oceanography, whose general objectives were "to assist in the development of the marine sciences, to encourage basic research and to advise the government agencies on various oceanographic problems." Thus "the problems of disposal of low level radioactive wastes" into ocean waters fit well within that committee's base of expertise and made it a logical choice for the AEC, Office of Naval Research, and Bureau of Commercial Fisheries to request investigation.[40] The AEC thus furthered environmental science research to help solve one of the challenges posed by nuclear research and development.

In the report, the NAS Committee on Oceanography attempted "to provide an estimate of the rate of return of radioactive substances to man, arising from stated rates of disposal into the coastal areas," which emphasizes a perceived connectedness between human and environmental health. The NAS committee said that the current practice of mixing low-level wastes with concrete and storing it in a 55-gallon drum would only provide containment for about ten years, but this should be long enough for the products to lose all radioactivity. No matter how safe the practice, the report declared that some sites would not be suitable (such as coastal estuaries, bays, and regions immediately seaward of these areas) and recommended more studies of coastal circulation and especially circulation of bottom waters. All in all, it claimed the dumping practice should be safe, unless shellfisheries were nearby, because radioactivity very possibly could sink into bottom sediments, get taken in by shellfish, and then consumed by humans. The authors considered this potential radioactive shellfish problem as the most serious danger of the radioactive dumping, again not because of the damage the shellfish and their ecosystems might incur but because humans might eat some of them. No matter the risks involved, as its final recommendation the report suggested, "The panel is of the opinion that certain Atlantic and Gulf of Mexico coastal areas can be used as receiving waters for the controlled disposal of packaged, low level, radioactive wastes."[41]

More interesting, the ways the report went into a great level of detail reflected a different position than decision makers had taken previously—nuclear waste became a bigger problem with each passing year due to increasing peacetime nuclear production, and the resulting wastes could not be disposed of by conventional methods (municipal incinerators, sanitary landfills, etc.). In all, the AEC dumped less than 6,000 curies of products between 1951–1958 in Atlantic waters, mostly in the form of "solid materials such as paper wipes, rags, mops, ashes, animal carcasses and contaminated laboratory paraphernalia."[42] Of course, playing back to the point that disposing of nuclear waste was fundamentally different than conventional wastes, the report pointed out that the type of isotope being dumped (e.g., strontium v. something less harmful like tritium) played an immense role in the environmental effects. Moreover, putting these products in the ocean differed a great deal from storing other wastes in landfills. For example, at a depth of 1,000 fathoms (a little over a mile), disposal canisters encountered over 3,000 pounds per square inch of pressure, and any rupture of those vessels would release radioactive products into the natural circulation of ocean waters. This would dilute the radioactive waste but also allow it to enter ecosystems. Apart from such known factors, a great many unknown issues—absorption factors and previously mentioned ocean circulation patterns—also played important roles. Therefore, no matter the environmental and scientific knowledge accumulated, the report summarized, "A precise evaluation of the quantity of radioactive substances that will be returned to man as the result of a stated rate of disposal into any one of the selected areas cannot be given."[43] Nevertheless, this did not stop the NAS committee from making produmping recommendations and policymakers from making decisions that implicitly assumed that they knew enough to go forward with dumping plans.

After dealing with nuclear waste for over a decade, US lawmakers worked to decrease federal responsibility for certain nuclear wastes. In September 1959, President Eisenhower signed Public Law (P.L.) 86–373 as an amendment to the Atomic Energy Act. The intention behind P.L. 86–373 was to allow the AEC to shift authority for disposing of nuclear waste over to state control, so long as the nuclear materials were "of less than a critical mass" (that is to say, could not be used to make a nuclear bomb). If a state's governor agreed and the AEC thought the state had an adequate program to deal with such wastes, the AEC would delegate some of its responsibility. Moreover, the AEC wanted the states to get behind P.L. 86–373 as quickly as possible, and as quickly as state regulatory programs could be "designed to protect the health and safety of the people against radiation hazards and to encourage the constructive uses of radiation." In doing so, the commission approved that when disposing of such radioactive materials, "certain limited quantities may be safely discharged into the air, water, and sewers, and buried in the soil."[44] This law was all part of normalizing nuclear waste products and likely as much about reducing public fears as helping the AEC reduce its workload. With this move, the commission showed that if it was willing for the states to handle such products, surely these wastes could not be terribly dangerous or worth much anxiety.

No matter how hard the AEC worked to dispel worries about disposal, its policies still could engender great fear, such as the minor international incident generated in late 1959 when the United States proposed granting a license to dump radioactive waste in the Gulf of Mexico. The license would let the waste be placed in the ocean equidistant from both US and Mexican territories, and the Mexican Embassy at Washington "expressed its opposition to the proposed license for unspecified scientific and technical reasons and for reasons of a political and public relations nature." Moreover, the Mexican government believed dumping so close to Mexican shores represented "a unilateral and arbitrary act on the part of the United States, any adverse results from which would present virtually identical hazards to the residents of the two countries." Although the US government had allowed them to attend the licensing hearings, Mexican officials declared their belief that if the situation had been reversed the United States would not feel it had received an appropriate say in the matter. Concerns also existed over why the selected site was

180 miles from both shores, "particularly as so little can be known with certainty in Mexico regarding the possible adverse effects oceanic waste disposal might have over a long period of time."[45]

Mexican protestations had their intended effect. A later AEC memo decreed that the United States should deny the dumping license because of the potential adverse effect on foreign relations with Mexico.[46] Early in 1960, a White House memo declared, "As a result of protests from Mexico, backed by the Department of State, the Atomic Energy Commission is considering denying a license for disposal of radioactive industrial wastes in the Gulf of Mexico."[47] Clearly, even though the AEC tried to mitigate any worries about disposal plans, nuclear waste dumping still engendered palpable fear among many, both in and out of the United States.

Moreover, the AEC knew that the public frequently disapproved of dumping decisions and intentionally tried to mask these as much as possible. A mid-December 1959 meeting of the AEC commissioners discussed the establishment of land disposal sites for radioactive wastes. At that meeting, the commissioners approved creating permanent land disposal sites on government-owned land (either federal or state) and authorized Oak Ridge in Tennessee and a site in Idaho as interim disposal sites, pending study and evaluation of other sites. However, in studying and approving those other sites, the commissioners declared at their meeting, "In accordance with past AEC practices when site selections were being made, site selection activities will be conducted with as little publicity as possible but that appropriate and useful public relations activities will be undertaken at the time of selection of sites to help assure public acceptance."[48] In very open language, then, the AEC's top policymakers agreed that plans for creating nuclear waste grounds should be withheld from the public whenever possible, with only very certain types of public relations spin even attempted. Whether decision makers truly thought that no problems existed with their dumping plans or not, they certainly knew that the general US public certainly would have significant concerns.

The AEC did attempt to dispel concerns when it could, however, such as when commissioners met with representatives from the State of New Jersey in early 1960. Particularly, the Garden State's delegates cited the previously mentioned National Academy of Sciences study on "Radioactive Waste Disposal into Atlantic and Gulf Coast Waters," which had indicated the possibility of inshore dumping. Even though the New Jersey Department of Health representative claimed that he knew of no health problems with any of the present disposal sites, the state's representatives remained apprehensive. The AEC responded, contrary to the NAS report's claims, that while the commission had plans for disposing wastes 150–230 miles off of Sandy Hook, New Jersey, it had no plans for inshore sites off the New Jersey coast. Moreover, both the AEC and New Jersey representatives realized that if the state officials had such problems with potential inshore disposal sites, then they "would have a major public relations problem in convincing the public that chemical processing plants handling significant quantities of radioactive materials could be operated safely within the state."[49] Worries about ocean disposal did begin to cause changes in decision-making.

The AEC of 1960 began to consider whether ocean dumping should be eschewed in favor of land disposal. One study showed that in most cases, land disposal "would be both feasible and less expensive than sea disposal." Reports claimed that if the AEC had pursued such a plan at the time, the temporary sites at Oak Ridge and in Idaho would be capable of handling all low-level radioactive waste produced by the United States until 1965. That study only focused on low-level wastes, however, since transportation costs for such were inexpensive, because those nuclear products required no special shielding. AEC chair John McCone did ask whether there was a "danger of buried waste material leaching radioactivity into the earth and eventually reaching rivers and streams," which demonstrated an understanding of how the natural world and its systems work. On top of the cheaper cost, however, "the risk of accidental release from the burial ground would not be significantly increased by burying a large amount of waste since there is adequate control of the burial ground."[50] Reevaluations of technical waste disposal matters continued as Eisenhower's presidency came to a close.

In late 1960, the commissioners discussed a letter to the Earth Sciences Division of the National Academy of Sciences–National Research Council. They intended their letter to reply to concerns held by the Earth Sciences Division about waste disposal, but internal discussion emphasized that the commissioners believed the NAS committee only held competence on geological aspects, and any discussion about waste disposal more generally was outside its field of knowledge. Moreover, the commissioners decided that their letter to the Earth Sciences Division should say, "However, we assume you do not mean that zero radioactivity should be allowed to reach man's environment. This would raise fundamental questions including those of a biological and medical nature that are very broad."[51] This response to members of the scientific community showed that scientific advice still had to be fitted into political realities, just as Allison Macfarlane argued at the beginning of this chapter.[52] The AEC letter also illuminated an assumption that there was nothing inherently wrong with releasing radiation into the natural world, so long as it was done in a controlled manner. The AEC thus recognized that its actions toed the line between waste disposal and controlled pollution.

The AEC report to Congress for 1959 provides a good endpoint for understanding AEC opinions on dumping during the Truman and Eisenhower presidencies. That report contained over seventy pages (nearly a fifth of the document) on nuclear waste disposal and comprised the most expansive treatise on the commission's positions and activities on the subject to date. The section claimed, "The major objective of waste management in atomic energy operations is control over the radiation hazard that might be produced by these wastes, either in storage or in nature." To this end, two basic disposal concepts existed—either concentrate wastes so these could be contained or dilute wastes so these could be dispersed. The section then proceeded to describe the "waste management methods" at several different AEC installations—a nuclear power plant (the Shippingport, Pa., Atomic Power Plant), a production and processing installation (Hanford Works, Hanford, Wash.), a development laboratory (Brookhaven, NY), and also disposal methods by sea and land burial.[53] Examining these three facilities individually sheds light on the overall thought patterns of the AEC.

The Shippingport facility, located on the Ohio River in Pennsylvania, is credited as the world's first nuclear power plant devoted solely to peaceful production of atomic electricity. The 400-acre site used pressurized water as a reactor coolant and in the process built up low-level radioactive waste from both corrosive processes and from fission products produced by occasional fuel ruptures. It also produced high-level wastes from the actual nuclear fuel. High-level wastes were shipped to an unnamed AEC site, and low-level wastes got reduced in concentration and discharged directly into the Ohio River, supposedly not to exceed one-tenth the maximum possible concentration. To ensure that the radioactivity of these lower-level wastes did not excessively pollute the river, the facility stored the liquids in large underground tanks for around forty-five days. The total "reactor-waste effluent" was about 23,000 gallons per month with radiation around 3 microcuries per milliliter (a little over 11 curies a month). In 1956, the commission began an off-site monitoring program for changes in air, soil, and vegetation in the area and also monitored well water within a mile of the site.[54] As at other sites, whenever possible the AEC discharged radioactive products directly into the local environment and, when this was not possible, stored that waste until such a time as it could be directly deposited into the environs, even if that day would never come.

The Hanford Site was 650 square miles in 1960, "located in a saucer-like basin surrounded by hills and mountains up to 3,600 feet above sea level" on the Columbia River in southeastern Washington.[55] One press release described Hanford as "constructed in this isolated expanse of wasteland" (yet, on the next page it discussed the people who lived near the plant).[56] There is a plateau in the basin where most of the plant is located, and the semiarid area was lightly populated at the time. A good thing, too, because as of January 1960 the Hanford plant had "discharged to the environment about 95 % of all low- and intermediate-level radioactivity so disposed of in the United States through atomic energy operations," making it a natural choice for study in the report. Disposal techniques used there depended on the site's unique location and geography and consequently would not necessarily work elsewhere. Eight reactors at Hanford produced plutonium for nuclear weapons, and those reactors had to be cooled by water from the Columbia River, which became contaminated by ambient radiation in the reactors. Prior to being released back into the Columbia River, the facility held the cooling water in tanks for one to three hours, which reduced the radioactivity by 50–70 percent. The report claimed, "By the time the effluent has traveled to the vicinity of Pasco, 35 miles downstream, and the first point of substantial use, further radioactive decay has reduced the gross activity by a total of about 90 percent and well below the permissible limits for safe consumption." Since the dilution of the Columbia is over 1.4 million gallons per second at places, this is unsurprising, but as chapter 2 demonstrated, determining "permissible limits" could be quite difficult and imprecise. Low-level cooling water with only minor radioactivity accounted for 30 billion gallons of the total waste created, but other waste existed with potentially far more harmful effects.[57]

To understand Hanford well, however, it must be situated within an environmental context. In August 1958, Donald A. Pugnetti, managing editor of the *Tri-City Herald* ("The Voice of Southeastern Washington"), sent a letter to White House Press Secretary James Hagerty. The letter itself seems fairly inconsequential in hindsight, but worth noting is that it was printed on the back of a hand-drawn map of the area surrounding the Hanford Atomic Works plant (see fig. 5.1).[58] That map situated the Hanford plant in the midst of a variety of both industrial centers as well as natural and agricultural elements— the Columbia and Snake Rivers, wheat farms, an apple orchard, and so forth.

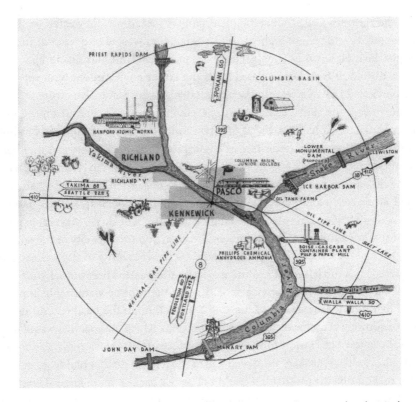

Fig. 5.1. Hand-drawn map of Hanford Atomic Works Area. Notice how natural and agricultural factors are interspersed with industrial centers of both traditional and nuclear varieties. DDEL, White House Central Files, General File, Box 1214, Folder 155, 1958, Letter from Donald A. Pugnetti to James Hagerty, August 16, 1958.

While it is possible that Pugnetti merely found a piece of scrap paper on which to type his letter, it seems more likely that the graphical representation meant something to him and helped convey a sense of how he thought about the area, especially the relationship between the Hanford Atomic Works plant and its surrounding environs. While the AEC studied Hanford and its radioactive waste disposal in terms of its distance from population centers, like Pasco, Hanford actually was part of a complex ecosystem dotted with mixed use agriculture and industry. To wit, studies of the effects Hanford had on local fish and wildlife began in the late 1940s and continued well into the 1950s.[59] Like any nuclear site, Hanford and its dumping could not truly be separated from its environment, both for good and bad.

The Hanford Site produced a great many other radioactive wastes, some of them solids and others highly toxic liquids. Solid wastes like "contaminated paper, boards, worn out tools, construction items, and aluminum spacers" were buried in trenches, isolated from the larger environment with very little perceived risk of affecting the water table. Bigger solid items were buried in very deep pits or stored in large concrete-lined tunnels. Perhaps more important, Hanford had produced fifty-two million gallons of high-level radioactive waste, stored indefinitely in underground tanks of between one-half to one-million-gallon capacity. The report asserted, "No environmental hazard exists as long as the tanks maintain their integrity." The site also had created around three billion gallons of intermediate-level waste, "deposited to the ground under carefully controlled conditions." For these, "favorable geological and hydrological conditions in the area, and the capacity of the soil to absorb isotopes, make it possible to hold the vast majority of the radioactive materials in a thick layer of sediments. Thus, the wastes are essentially 'stored' in the ground, and any water percolating through to the water table is purified by time and the action of the soil."[60] Even though the local environmental conditions may have mediated the ways in which radioactive waste disposal occurred at Hanford, the site still suffered from the same problems as anywhere else—the AEC produced a great deal of waste that had to go somewhere, and this meant that if it could not be put back into the natural world it had to be stored until such a time came (or never came) that the waste could be safely put back into the environment.[61]

Finally, the report surveyed waste disposal at Brookhaven National Laboratory, a 3,600-acre site located at the center of Long Island, New York, devoted to nuclear research. Most of the laboratory's waste came either directly or indirectly from the large air-cooled research reactor on-site. "The off-gases from the hot laboratory are cleaned by filters and scrubbers and released through a pipe going up the center of the 310-foot stack provided for the reactor cooling air." Radioactive argon-41 was the most significant radioactive product in the cooling air, but the stack spit the gas up very high, where presumably the radioactivity would not affect humans and could be diluted by the general air. Any liquid wastes were of a low level and "discharged to a sewerage system installed when the site was used for a large Army camp. The effluent passes through an Imhoff tank that removes most of the solids and then is discharged to a large sand filter, collected by an underlying tile field, chlorinated, and discharged to a small stream."[62] Like at other sites, the AEC produced significant radioactive waste and thus had to diffuse that radiation into the natural world, planned as carefully as possible so that such dispersal hopefully would not affect humans.

Apart from these three sites, more general waste disposal occurred both by land and sea burial. The report for 1959 stated, "Except for storage in rigorously maintained tanks, there is no waste management method that withholds radioactive wastes from the environment on an essentially permanent basis." However, the AEC did not consider indefinite storage necessary for most wastes, as these would lose all radioactivity in a few years. "Land and sea burial are means of disposal intermediate between long-term storage and diluted release to the environment," even if only low-level wastes could be buried in the ocean. Brookhaven, for example, buried radioactive wastes in the ocean. The laboratory reduced some of its wastes to solid form, sometimes with a solidifying agent like concrete, and from there essentially stored the waste into fifty-five gallon drums, had trucks drive these to a wharf where the barrels could be loaded onto ships and taken out 250 miles to sea, and dumped the barrels overboard at depths of more than a mile. While other countries may have dumped liquid wastes directly into the sea, the AEC did not license such practices (even if waste could be unloaded directly into streams and rivers that flowed to the ocean). All told, tens of thousands of drums and hundreds of concrete boxes found a watery resting place this way.[63]

Even with such well-established programs, the AEC did not find its waste disposal program sufficient. The commission therefore enacted an "extensive, coordinated research and development program in all phases of waste control operations." That research program's three objectives were: "develop practical systems for the final disposal, or long-term management, of highly radioactive wastes associated primarily with the chemical reprocessing of irradiated nuclear fuels"; "evaluate quantitatively the dilution or concentration factors in nature in order to determine the degree of treatment required prior to release of low-level wastes to the atmosphere, ground, or waterways"; and "obtain increased knowledge of the fundamental phenomena and processes involved in handling and disposal of radioactive wastes so that more efficient and economical systems may be devised." The AEC recognized that high-level wastes would plague humans for hundreds of years and thus needed to "be contained essentially at the point of disposal so that man, his environment, and his resources are not adversely affected." Tank storage may have worked for fifteen years but obviously did not represent true disposal "in the ultimate sense." The commission also recognized that its knowledge of "oceanic behavior" was inadequate, "and attendant engineering problems appear so complex for high-level wastes, that alternate systems that are easier to control directly are the most likely solution." Other research existed as well into every different type of wastes at all radioactivity levels.[64]

In sum, the 1959 AEC report to Congress showed two things—by the end of the 1950s the commission greatly cared about how disposal happened and sought to improve that, and yet, for all the research and care paid to such matters, it still lacked practical solutions for some wastes with the realization that such answers might never exist. The AEC report to Congress for 1960 reported that, by 1959, the United States had produced sixty-five million gallons of high-level wastes.[65] Considering that no solution yet existed for high-level wastes other than storage, this represented an incredible liability for both the AEC and the United States. Moreover, those many millions of gallons of deadly sludge serve as a fitting symbol for the conundrum that is nuclear waste disposal. Many fantastic and incredible feats can be accomplished through nuclear technology, but almost all of them produce nuclear wastes that range from mildly dangerous to horribly toxic. Once the wastes are produced, there is nothing humans can do to reduce the inherent radiation other than wait for it to decay over many half-lives, some of which can take much longer than any human life.

It is fitting, then, if incredibly unfortunate, that the legacy of nuclear waste disposal policies has outlived essentially every policymaker from the early Cold War era. In February 1986, the US public learned through a Freedom of Information Act request that the Hanford facilities in southeastern Washington State had pumped millions of curies of radiation into the local environment (for comparison, the supposed disaster at Three Mile Island amounted to a paltry fourteen curies). The 1949 "Green Run" at Hanford, for example, saw the release of several thousand curies of Iodine-131 into the atmosphere to test the environmental effects. The experiment originated from a desire to improve environmental science enough that it could strengthen national security through better monitoring systems of the Soviet plutonium production plant at Mayak. But weather and precipitation destroyed the scientific value of the experiment and rained radioactive residue on crops and downwind communities (which included Spokane, 125 miles from the release point).[66] Hanford was created to help produce the plutonium deemed necessary for the atomic bombs used to win World War II and the Cold War—to safeguard the United States and its people. At times, nuclear technologies clearly had the opposite effect. And in a situation reminiscent of the decisions of AEC commissioners, Hanford's top decision makers sometimes intentionally sacrificed human safety for the sake of easier public relations.[67]

Historian Andrew Jenks has argued that, at another site, the Lake Ontario Ordnance Works in western New York State, "a spirit of patriotic sacrifice, combined with a culture of secrecy, overrode concerns about safety." The site was a TNT plant during World War II, converted to a radioactive dumping ground in 1944, and variously sold back in pieces to both public and private entities in the 1950s and 1960s. All the while, contamination by nuclear waste created incredible human and environmental health dangers that last even through current-day cleanup operations.[68] And at the planned Yucca Mountain nuclear waste repository, designed as the ultimate nuclear waste disposal site for the United States, original plans from the 1980s have been revised to include a "titanium drip shield" because of the higher amounts of water than initially believed. Water would corrode waste disposal containers, which necessitates the shield. One opinion piece in the *Bulletin of the Atomic Scientists* has doubted whether such a shield could be installed due to both practical and political concerns.[69] Additionally, an explosion occurred on Valentine's Day 2014 at the New Mexican Waste Isolation Pilot Plant (WIPP) because contractors accidentally packed low-level nuclear wastes with the wrong type of cat litter (wheat instead of clay based).[70] Disposing of Cold War–era nuclear waste remains a current day, unresolved problem.

Typically, waste disposal in the United States, during both the early Cold War and current times, has produced an "out of sight, out of mind" mentality among US citizens and policymakers.[71] (In analyzing the abundant nuclear images in US society, historian Spencer Weart has even argued that radioactive waste from nuclear power plants has frequently been compared to regular sewage or even human excrement.[72]) But an "out of sight, out of mind" mindset does not make sense when applied to nuclear waste (and likely other wastes as well). The country's top decision makers relied on scientific knowledge to deal with the sizable problem of disposing of intentionally produced radioactive waste (let alone dealing with unexpected, true nuclear disasters like at Chernobyl and Fukushima).[73] Tens of billions of gallons of low-level waste represented no significant technical obstacle but still polluted the nation's rivers and soils and the global oceans. The nation only had hopes and assurances (not all born out or true) that such contamination would not permanently harm US environments or peoples. The other tens of millions of gallons of high-level contaminated waste could be stored more or less safely for a time in gigantic vats but required complete removal from the environs, or else great biological devastation would occur. Previous ways of thinking about trash may have worked for previous types of trash (or at least caused less acute problems), but such approaches were less successful when applied to nuclear waste. The AEC consistently debated the limits to how much radiation could and could not be placed into the natural world and used environmental science to better understand how safely to do so. But perhaps a better way of thinking would have been to question whether any extra radiation should be permitted into the lands and bodies of US peoples. US policymakers could have reoriented their thinking from acceptable limits toward a position that declared no extra radiation was acceptable. But doing so would have necessitated reconsidering the activities that produced such radiation, something the AEC was not willing to do.

Conclusion

On Nuclear Technologies, Decision Making, and Environmentalism

In historian Donald Worster's estimation, the July 1945 Trinity detonation of the world's first atomic bomb in the New Mexico desert marked the beginning of the "Age of Ecology." This declaration is partially ironic. Worster believed that the overwhelming destructive power of nuclear weapons awed and humbled people into thinking about their place in the world. He wrote that, in addition to radioactive fallout that threatened human and environmental health, another "kind of fallout from the atomic bomb was the beginning of widespread, popular ecological concern around the globe."[1] More than just the fear of environmental contamination, however, the nuclear age also helped birth ecology by the use of radioisotope tracers being injected into various ecosystems to follow their path through various trophic flows in the research of scientist Eugene Odum, among others.[2] Such interaction was not unidirectional, however.

Not only did the advent of the nuclear age help to give rise to ecology as a coherent discipline but environmental science more broadly became intermingled into the development of nuclear technologies. By the time President Dwight Eisenhower left office, executive decision makers in the United States had, by and large, come to depend on environmental science to craft their nuclear policies. In a basic sense, they employed scientific knowledge about the natural world to develop and implement nuclear technologies. But, on a more fundamental level, US policymakers utilized their environmental understandings to support larger goals for nuclear technologies—protect the United States, improve the lives of its citizens, and advance the nation in a Cold War geopolitical context. In doing so, we see a broader story of the inclusion of environmental science into Cold War policymaking about nuclear technologies: the individual policymakers themselves often feature as less important than the agencies in which they worked, as policy goals that privileged US strategic interests, secrecy, and nuclear boosterism consistently won out over environmental and health considerations no matter who held the positions in either the Truman or Eisenhower administrations.

Such an interaction between environmental science and policymaking about nuclear technologies happened in myriad ways. During nuclear tests, meteorology proved influential in determining when tests occurred and what happened when these did. Nuclear weapons also could have incredible effects on the natural world, and studies of bomb test sites and their surrounding environs reveal an increasing reliance on environmental science to understand the applications of nuclear science. After tests, scientists gained an evolving knowledge of the radioactive fallout produced, and the ways decision makers used these scientific understandings of fallout reflect a developing but innate understanding of an interconnection between human health and the environment. That comprehension proved crucial in the later years of Eisenhower's presidency when he and others sought to end nuclear weapons tests, as concern for the natural world and knowledge of it became critical points of contention in nuclear test ban talks. Atomic agriculture and nuclear waste disposal provide further evidence for the interconnected nature of decision-making, environmental science, and related technologies. We cannot understand how nuclear science and its applications developed during the early Cold War without understanding the reciprocal relationship between nuclear and environmental science.

For all that knowledge, however, little-to-nothing about the US nuclear program benefited the natural world. Instead, geopolitical and natural security priorities caused top decision makers to weigh their concerns for the nation and decide what they thought was in the country's best interest. Sometimes protecting the nation meant developing natural resources or preserving human health by keeping the environment free from nuclear pollutants. More often, however, fear of Soviet military aggression meant such environmental concerns were sacrificed in the name of national security. Moreover, many examples throughout the book of how nuclear waste dumping or radioactive fallout affected non–US peoples clearly demonstrate that US policymakers almost always cared overwhelmingly about the United States and its environments (or, more likely US citizens, residents, and voters). For example, since tropospheric fallout tends to stay in similar latitudes, tropical peoples suffered disproportionately from the fallout produced by nuclear tests at the Pacific Proving Grounds deemed necessary for US national security. Biology, geology, ecology, and the like found useful implementation for what those could do to help develop nuclear technologies and not the natural world itself. Nobody would bat an eyelash at a farmer utilizing chemistry or soil science, nor an oil wildcatter using seismology. Yet executive policymaking concerning nuclear technologies contained a similar reciprocal relationship between decision making and environmental science not often seen to such a degree in other fields.

While this book has specifically concerned itself with the ways early Cold War leaders in the United States incorporated environmental science into their policymaking about nuclear technologies, in many ways it has also sought to comment more broadly on the relationship between political leaders and the natural world writ large.[3] A certain irony exists in the incorporation of environmental science into the administration of nuclear technologies. While nuclear technologies heavily depended on environmental science to develop, in doing so those same technologies frequently caused great harm to the natural world.[4] Nuclear technologies threatened the very entities that generated so much data and information necessary for the advancement of nuclear science. In terms of administrating the research and development of those technologies, while executive decision makers steeped themselves in environmental knowledge, caring deeply about what they could learn from the natural world, they frequently made choices that proved catastrophic for local and global environments. In one of his more sanguine moments, Worster explained his hope that the discipline of ecology, which he saw as having devolved into seeing most or all ecological change rooted in disorder and chaos, as potentially offering "a pathway to a kind of moral enlightenment that we can call, for the purposes of simplicity, 'conservation.'"[5] Clearly, however, the desire to cultivate and improve environmental knowledge is something very different than environmentalism. Defining environmentalism, however, can be tricky.

"Environmentalism," historian Hal Rothman contended, "is one of the most important new dimensions to appear in American society in the post–1945 world. Part social movement, part manifestation of the increasing affluence and privilege of American society and different from the conservation movement that preceded it, environmentalism took center stage in the transformation of the values and mores of the second half of the twentieth century."[6] Even though environmentalism has clearly played an important role in postwar United States society, both popular and scholarly understandings of the idea are imperfect. Cultural scholar Raymond Williams defined environmentalism as "concern with the human and natural habitat" or "the doctrine of the influence of physical surroundings on development."[7] Yet more common definitions eschew such a value-neutral characterization to emphasize protecting the environment at the expense of other considerations, most especially economic concerns.[8]

Environmentalist-like care for the environment in the United States was nothing new by the mid-twentieth century. In 1864, George Perkins Marsh wrote *Man and Nature* not only to point out how much humans could effect change in the natural world, but also "to point out the dangers of imprudence and the necessity of caution in all operations which, on a large scale, interfere with the spontaneous arrangements of the organic or the inorganic world."[9] After Marsh, explicit concern for the natural world increased dramatically from late-nineteenth century efforts to save the American Bison through Progressive-Era conservationism.[10] By 1949, just a few years after the Trinity test perhaps birthed the "Age of Ecology," famed forester and environmental thinker Aldo Leopold's highly influential *A Sand County Almanac* urged readers to "think like a mountain."[11] And even though recent scholarship has questioned whether it was in actuality a radical text, by many traditional accounts the environmental age truly began sometime around the 1962 publication of Rachel Carson's *Silent Spring*.[12] The text served as a call to action for many budding environmentalists of the 1960s.[13] Or, perhaps it was the first Earth Day during the spring of 1970 that truly marked the advent of a coherent, coalesced environmentalist movement.[14] The hefty historiography surrounding the organic development of environmentalism over perhaps more than a century suggests that pinning down when environmentalism became an intellectually robust idea is quite difficult.

Regardless of when environmentalism developed, it is fair to ask what we can reasonably expect of our elected officials in terms of environmentalist-like care for the natural world. Historian Otis Graham has argued that the nation should look "for a leader who has slept under the stars" or spent time as a youth out in nature. He further elaborated that when "recruiting our governing elites, we should give the edge to those who have managed sizable institutions where science is respected."[15] And yet, this book has shown that both Truman and Eisenhower did just that in the White House. Moreover, not only did they respect science but the two presidents depended on environmental science and presided over various institutions that did the same, at least when administrating nuclear technologies. Their presidential administrations show an ever-increasing appreciation for, if not the natural world itself, then at least how scientific knowledge of the environment could make the development of nuclear technologies possible. While deep down in our guts we may have the feeling that our political leaders would make decisions that better protected environmental health if they only had enough knowledge about how their decisions affected (most often hurt) the natural world, to the contrary, scientific environmental knowledge alone clearly cannot engender care for the natural world.[16]

On the other hand, perhaps it is nuclear technologies that confound this entire discussion. Sixty years after the detonation of the first atomic bomb, historian Andrew Bacevich declared, "More than America's matchless material abundance or even the diffusion of pop culture, the nation's arsenal of high-tech weaponry and the soldiers who employ that arsenal have come to signify who we are and what we stand for."[17] If one were to believe Bacevich, nuclear weapons would then seem to characterize the nation's culture more than any other technology under its control. And yet, historian Lawrence Keeley has argued that after World War II, the atomic bomb's mushroom cloud symbolized the "newly discovered madness of war."[18] The two statements are not necessarily antithetical—nuclear weapons have occupied a complicated place in the United States since humans first harnessed the power of the atom. From the outset, the bomb and other nuclear technologies could be anything from guardian angel to demonic terror, depending on one's point of view.

More recently, concurrent with President Barack Obama's historic May 2016 visit to Hiroshima, Japan, the *Wall Street Journal* conducted a public opinion poll on atomic bomb use. In a September 1945 poll by Roper, 78 percent of respondents had believed that the United States dropping two atomic bombs on Japan was either just right or even not enough. When the *Wall Street Journal* tried to replicate that poll in 2016, that number had fallen to only 31 percent. This 47 percent plunge in willingness by the public to drop nuclear bombs on Japan might be understood as a US public that has become much more concerned about the moral, political, and environmental consequences of using nuclear weapons in war. However, when presented with a hypothetical yet similar situation of needing "to reach Tehran and force the Iranian government to capitulate (at an estimated cost of 20,000 American fatalities), or shock Iran into unconditional surrender by dropping a single nuclear weapon on a major city near Tehran killing an estimated 100,000 Iranian civilians (similar to the immediate death toll in Hiroshima)," 59% supported using the bomb.[19] Over the past seven decades, perhaps the public has changed its opinions on Japan rather than become wary of nuclear weapons.

No matter the complicating factors, we must come to grips with the notion that it will take more than mere environmental knowledge for politicians and leaders to take protecting the natural world seriously. It will even take more than attentiveness to the true effects of technology on the natural world. Nuclear technologies developed in explicit dialogue with environmental science, and executive policymakers had a good grasp of how their policy decisions would affect global environments, but that clearly did not save global peoples and environments from the ravages of those technologies and their crippling radiation. While executive policymakers sometimes expressed real anxieties about the environmental damage wrought by nuclear technologies, they much more often made choices that privileged nuclear boosterism and secrecy. The single biggest mistake policymakers made, then, is that they respected institutional priorities over the lives those institutions were ostensibly charged to protect and enrich. The decision makers in this book took seriously their charge to protect and advance the welfare of the United States and its people, but they often failed to do so because of their loyalties to nuclear technologies over human bodies.

Thinking more broadly, historian Etienne Benson succinctly wrote in his book *Surroundings*, "There have been many ways of being environmental since the emergence of the concept sometime between the late eighteenth century and the mid-nineteenth century; that particular ways of being environmental have emerged to serve particular aims under particular circumstances; that while none of these ways are either illegitimate or perfect, some of them are no longer very well suited to present-day aims and circumstances; and that we will as a consequence almost certainly need new ways of conceiving and relating to our environments in the future, for which the past may serve as guide."[20] Too true. Recognizing that it will take more than scientific knowledge about the environment to protect the natural world and human bodies innately connected to it, the events studied in this book can help demonstrate that what is needed are policymakers who value the human beings in this country more than the institutions they administrate.

Such a mindset need not be an inherently partisan issue, either. While in recent years many liberals have taken up environmentalism as a key liberal issue, in 2012, philosopher Roger V. Scruton argued that conservatives represented the nation's true environmentalists and thus needed to resurrect their environmentalist credentials and reclaim that moniker from liberals.[21] This author would argue that any politician who truly cares about the welfare of US citizens, no matter their political leanings, should be concerned with protecting the environments in which those citizens live. Every voter, no matter their political leanings, still has to live on the same planet.

To that point, in a 1953 commencement address at Seattle University, Thomas E. Murray, Commissioner of the Atomic Energy Commission, reminded those gathered that we are creatures inherently connected to the natural world around us.

> Tornado and earthquake and resulting conflagration have long been called acts of God because man, humble in the consciousness of his own limitations, has until this decade recognized them as beyond his capacity to produce or control. . . . Today man is more powerful. He can, as it were, generate hurricanes, earthquakes, and consuming fires. He can today open the tight doors of the atom and let forth all three—wind, earthquake, and fire— in such a manner as to make Hiroshima's atomic attack look like a Civil War bombardment. Because of the limitless nature of our destructive potential power we must moderate our forceful capabilities with something of the meekness and patience of the Saints. We must learn something of God's contempt for the great and the mighty—something of His preference for His little ones.[22]

In such a worldview, it would be tough to argue with Murray. If he was correct that humans were in control of the natural world with "limitless" power, then humans would indeed have an awesome responsibility to use that power wisely and judiciously or perhaps not at all.[23]

The combination of nuclear technologies and executive decision making from 1945–1960 provides numerous examples of the importance of environmental knowledge to politicians and bureaucrats from varied political perspectives. We could perhaps interpret the inculcation of environmental science into policymaking as the natural world being too vast, too omnipresent in our lives for humans to ever leave it out of our considerations and decisions. Or we can believe, as historian Matthew Booker has counseled, "While the past entraps it can also liberate; it can remind us of possibilities we did not know we had."[24] In this way, we should look to early Cold War policymaking in the United States less to illuminate a fundamental truth about the natural world and more to comprehend the capacity of those in power to influence the lives of humans across the entire world. If we can learn from both the successes and failures of policymaking about nuclear technologies and environmental science, then perhaps we can better understand what we should expect of our leaders and how we should, as conscientious, thinking humans, interact with the environment when we build modern societies that benefit and protect everyone involved.

Notes

INTRODUCTION

1. Defense Technical Information Center (hereafter DTIC), "Operation UPSHOT-KNOTHOLE, 1963," Accession Number ADA121624, 100.

2. DTIC, "Operation UPSHOT-KNOTHOLE, 1963," 100.

3. As a US Army Major famously quipped in 1968 during the heart of the Vietnam War, "It became necessary to destroy the town to save it." "Major Describes Move," *New York Times*, 8 Feb. 1968, 14.

4. Ann Finkbeiner, "How Do We Know Nuclear Bombs Blow Down Forests? Because We Built a forest in Nevada and blew it down," *Slate*, 31 May 2013, http://www.slate.com. Accessed on 23 July 2013; "Atomic Bomb TREES," YouTube, http://www.youtube.com. Accessed on 5 July 2013.

5. See, for example, the section titled "Substrata of Environmental Concerns: The Nuclear Apocalypse and Cancer Fears" in Joachim Radkau, *Nature and Power: A Global History of the Environment*, trans. Thomas Dunlap (New York: Cambridge University Press, 2002, 2008), 265–72.

6. Novick continued, "We are slowly coming to realize that the thin film of life which covers the earth is a single complex web; a chord plucked at any one point vibrates throughout the world. And man is part of this web; he depends on it totally for his food, air, and water." Sheldon Novick, "The Menace of the Peaceful Atom," *Commentary*, 1 December 1968, 33–40. Accessed 18 May 2016 at https://www.commentary-magazine.com. The next year, Novick published *The Careless Atom* (Boston: Houghlin Mifflin, 1969). For a greater context on Novick, *Commentary*, and the rise of antienvironmentalism, see Alex Boynton, "Formulating an Anti-environmental Opposition: Neoconservative Intellectuals during the Environmental Decade," *The Sixties: A Journal of History, Politics and Culture* 8, no. 1 (June 2015): 1–26.

7. She continued, "Well why doesn't President Obama know this? He's an intelligent man. He's got two little girls he loves. What the hell does he think he's up to

supporting the nuclear industry? It's wicked." *Pandora's Promise*, directed by Robert Stone (2013; Robert Stone Productions, Vulcan Productions).

8. Ferenc M. Szasz, *The Day the Sun Rose Twice: The Story of the Trinity Nuclear Explosion, July 16, 1945* (Albuquerque: University of New Mexico Press, 1985), 5.

9. Though not at all in the way Worster intended, the Encore test's recreated forest space exemplified all three levels of analysis outlined here: Donald Worster, "Transformations of the Earth: Toward an Agroecological Perspective in History." *Journal of American History* 76, no. 4 (March 1990): 1090–91.

10. See, for example, chapter 6, "Ecology and the Atomic Age," in Joel B. Hagen, *An Entangled Bank: The Origins of Ecosystem Ecology* (New Brunswick: Rutgers University Press, 1992), 100–121. Hagen particularly traces the research of Eugene and Howard Thomas Odum who pioneered ecology as an academic discipline, especially by using radioisotopes for research. Similarly, see chapter 10, "Ecosystems," in Angela N. H. Creager, *Life Atomic: A History of Radioisotopes in Science and Medicine* (Chicago: University of Chicago Press, 2013); also Laura J. Martin, "Proving Grounds: Ecological Fieldwork in the Pacific and the Materialization of Ecosystems," *Environmental History* 23:3 (Jul. 2018): 567–92. The entire October 2003 issue of *Social Studies of Science* was devoted to a related idea. See John Cloud, "Introduction: Special Guest-Edited Issue on the Earth Sciences in the Cold War," *Social Studies of Science* 33, no. 5 (Oct. 2003): 629–33.

11. Mark D. Merlin and Ricardo M. Gonzalez, "Environmental Impacts of Nuclear Testing in Remote Oceania, 1946–1996," in *Environmental Histories of the Cold War*, J. R. McNeill and Corinna R. Unger, eds. (New York: Cambridge University Press, 2010), 167.

12. Mark Fiege, "The Atomic Scientists, the Sense of Wonder, and the Bomb," *Environmental History* 12, no. 3 (Jul. 2007): 584.

13. Paul S. Sutter, "The World with Us: The State of American Environmental History," *Journal of American History* 100, no. 1 (June 2013): 100.

14. Mark Fiege philosophically approached this idea in chapter 2 "By the Laws of Nature and Nature's God: Declaring American Independence" in his work *The Republic of Nature: An Environmental History of the United States* (Seattle: University of Washington Press, 2012, 2013), 57–99. See also Radkau, *Nature and Power*, xii. However, while policymakers did indeed involve the environment in their policy decisions concerning nuclear technologies, this is not synonymous with environmentalist policymaking. The question of environmental understanding leading to more environmentalist policymaking (or not) is briefly examined in the section "*New Awareness or Better Organization?*" in Radkau, *Nature and Power*, 324–26. Similarly, Richard White argued, "A connection with the land through work creates knowledge, but it does not necessarily grant protection to the land itself." Richard White, "'Are You an Environmentalist or Do You Work for a Living?': Work and Nature," in *Uncommon Ground: Rethinking the Human Place in Nature*, William Cronon, ed. (New York: W. W. Norton & Company, 1996), 181.

15. Jacob Darwin Hamblin, *Arming Mother Nature: The Birth of Catastrophic Environmentalism* (New York: Oxford University Press, 2013), 3–4.

16. Etienne Benson, *Surroundings: A History of Environments and Environmentalisms* (University of Chicago Press, 2020), especially pp. 7–9. Nature has become a loaded term in environmental history, hence not using it is as much a practical as intellectual decision. And yet, as historian Paul Sutter notes, while the term "nature" is "freighted and now thoroughly problematized," the word "environment" is "surprisingly undertheorized." Paul S. Sutter, "The World with Us: The State of American Environmental History," *Journal of American History* 100, no. 1 (June 2013): 97. On the subject and particularly influential to this work, see Julia Adeney Thomas, *Reconfiguring Modernity: Concepts of Nature in Japanese Political Ideology* (Berkeley: University of California Press, 2001), 2–3; William Cronon, "Foreword to the Paperback Edition," in *Uncommon Ground: Rethinking the Human Place in Nature*, William Cronon, ed. (New York: W. W. Norton & Company, 1995, 1996), 20; Raymond Williams, *Keywords: A Vocabulary of Culture and Society* (New York, NY: Oxford University Press, 1976, 1983), 219–24; Anne Whiston Spirn, "Constructing Nature: The Legacy of Frederick Law Olmstead," in *Uncommon Ground: Rethinking the Human Place in Nature*, ed. William Cronon (New York, NY: W. W. Norton & Company, Inc., 1995), 101; Paul S. Sutter, *Driven Wild: How the Fight against Automobiles Launched the Modern Wilderness Movement* (Seattle: University of Washington Press, 2002). There has been less written on "environment," but see Shannon O'Lear, *Environmental Politics: Scale and Power* (New York: Cambridge University Press, 2010), 6–7. On connections between the natural world and human bodies, which further complicate the idea that there might be a nature or environment separate from humans, see, for example, Julia Adeney Thomas, "History and Biology in the Anthropocene: Problems of Scale, Problems of Value," *American Historical Review* 119, no. 5 (Dec. 2014), 1592–97; Linda Nash, *Inescapable Ecologies: A History of Environment, Disease, and Knowledge* (Berkeley: University of California Press, 2006).

17. These are fairly standard political science definitions that can be found in many introductory government textbooks, for example, Karen O'Connor, Larry J. Sabato, and Alixandra B. Yanus, *American Government: Roots and Reform* (Upper Saddle River: Pearson Education, Inc., 2014), 251.

18. Brian Balogh, *Chain Reaction: Expert Debate and Public Participation in American Commercial Nuclear Power, 1945–1975* (Cambridge: Cambridge University Press, 1991), 16–17, 19–20, 62–66.

19. Karl Boyd Brooks, *Before Earth Day: The Origins of American Environmental Law, 1945–1970* (Lawrence: University Press of Kansas, 2009).

20. Hersey's work started as a single issue of the *New Yorker* magazine but soon became a monograph. John Hersey, *Hiroshima* (New York: Bantam Books, Inc., 1946, 1981). *Them!* was Warner Brothers' highest grossing film of 1954 and a bona fide box office hit. William M. Tsutsui, "Looking Straight at *Them!* Understanding the Big Bug Movies of the 1950s," *Environmental History* 12, no. 2 (April 2007): 237.

21. As more than a generation of scholarship has shown, the line between natural world and bodies, of course, proves difficult to pin down. Even so, the following works overwhelmingly were not environmental histories in a traditional sense. On bodies: Phillip L. Fradkin, *Fallout: An American Nuclear Tragedy* (Tucson: University of Arizona Press, 1989); Richard L. Miller, *Under the Cloud: The Decades of Nuclear Testing* (New York: The Free Press, 1986); Howard Ball, *Justice Downwind: America's Atomic Testing Program in the 1950s* (New York: Oxford University Press, 1986); Barton C. Hacker, *Elements of Controversy: The Atomic Energy Commission and Radiation Safety in Nuclear Weapons Testing, 1947–1974* (Berkeley: University of California Press, 1994); *Nuclear Wastelands: A Global Guide to Nuclear Weapons Production and Its Health and Environmental Effects*, Arjun Makhijani, Howard Hu, and Katherine Yih, eds. (Cambridge: MIT Press, 1995); On culture, society, and specific tests: Paul Boyer, *By the Bomb's Early Light: American Thought and Culture at the Dawn of the Atomic Age* (New York: Pantheon Books, 1985); Paul Boyer, *Fallout: A Historian Reflects on America's Half-Century Encounter with Nuclear Weapons* (Columbus: Ohio State University Press, 1998), xii; Spencer R. Weart, *The Rise of Nuclear Fear* (Cambridge: Harvard University Press, 2012); Debra Rosenthal, *At the Heart of the Bomb: The Dangerous Allure of Weapons Work* (Reading: Addison-Wesley Publishing, Inc., 1990); Robert A. Jacobs, *The Dragon's Tail: Americans Face the Atomic Age* (Amherst and Boston: University of Massachusetts Press, 2010); Szasz, *The Day the Sun Rose Twice*; J. Samuel Walker, *Prompt and Utter Destruction: Truman and the Use of Atomic Bombs Against Japan* (Chapel Hill: University of North Carolina Press, 1997, 2004); Kai Bird and Martin J. Sherwin, *American Prometheus: The Triumph and Tragedy of J. Robert Oppenheimer* (New York: Alfred A. Knopf, 2005); Dan O'Neill, *The Firecracker Boys* (New York: St. Martin's Press, 1994); Curtis Frederick Foxley, "The Business of Atomic War: The Military-Industrial Complex and the American West," PhD diss., University of Oklahoma, 2020; *The Atomic West*, Bruce Hevly and John M. Findlay, eds. (Seattle: University of Washington Press, 1998); Brian Balogh, *Chain Reaction*; Michael Bess, *The Light-Green Society: Ecology and Technological Modernity in France, 1960–2000* (Chicago: The University of Chicago Press, 2003); Gabrielle Hecht, *The Radiance of France: Nuclear Power and National Identity after World War II* (Cambridge: MIT Press, 1998, 2009).

22. The history of nuclear technologies can and should be understood as a broader intersection of militarization and technology, meaning that histories of war and environment are particularly important. The two seminal works here are Edmund Russell, *War and Nature: Fighting Humans and Insects with Chemicals from World War I to Silent Spring* (New York: Cambridge University Press, 2001) and *Natural Enemy, Natural Ally: Toward an Environmental History of War*, Richard P. Tucker and Edmund Russell, eds. (Corvallis, Oregon: Oregon State University Press, 2004). In that work, especially see William M. Tsutsui, "Landscapes in the Dark Valley: Toward an Environmental History of Wartime Japan." Also important: J. R. McNeill, "Woods and Warfare in World History," *Environmental History* 9, no. 3 (Jul. 2004); Lisa M. Brady, "The

Wilderness of War: Nature and Strategy in the American Civil War." *Environmental History* 10, no. 3 (Jul., 2005); Chris Pearson, *Scarred Landscapes: War and Nature in Vichy France* (New York: Palgrave Macmillan, 2008); Chris Pearson, Peter Coates, and Tim Cole, eds., *Militarized Landscapes: From Gettysburg to Salisbury Plain* (New York: Continuum, 2010); Lisa M. Brady, *War Upon the Land: Military Strategy and the Transformation of Southern Landscapes During the American Civil War* (Athens: University of Georgia Press, 2012); Edwin Martini, ed., *Proving Grounds: Weapons Testing, Militarized Landscapes, and the Environmental Impact of American Empire* (Seattle: University of Washington Press, 2015).

23. Jacob Darwin Hamblin, *Poison in the Well: Radioactive Waste in the Oceans at the Dawn of the Nuclear Age* (New Brunswick: Rutgers University Press, 2008), chapter 7; "Atomic Sublime: Toward a Natural History of the Bomb," in Fiege, *The Republic of Nature*, 281–317. See also Martin V. Melosi, *Atomic Age America* (Boston: Pearson, 2012).

24. Kate Brown, *Plutopia: Nuclear Families, Atomic Cities, and the Great Soviet and American Plutonium Disasters* (New York: Oxford University Press, 2012); Mark D. Merlin and Ricardo M. Gonzalez, "Environmental Impacts of Nuclear Testing in Remote Oceania, 1946–1996," in J. R. McNeill and Corinna R. Unger, eds., *Environmental Histories of the Cold War* (New York: Cambridge University Press, 2010), 167–202. See also Ralph H. Lutts, "Chemical Fallout: Rachel Carson's *Silent Spring*, Radioactive Fallout, and the Environmental Movement," *Environmental Review* 9, no. 3 (Autumn 1985).

25. Separating humans from our environments has become especially difficult with revelations about the human microbiome, especially the notion that one hundred trillion bacteria live in and on our bodies. Gina Kolata, "In Good Health? Thank Your 100 Trillion Bacteria," *New York Times*, 13 June 2012. Also, on technology and society, in addition to the previously mentioned Ellul, *The Technological Society*, see: Lewis Mumford, *Technics and Civilization* (New York: Harcourt, Brace and Company, 1934); Rosalind Williams, "Classics Revisited: Lewis Mumford's *Technics and Civilization*," *Technology and Culture* 43, no. 1 (Jan., 2002); Langdon Winner, *Autonomous Technology: Technics-Out-of-Control as a Theme in Political Thought* (Cambridge: MIT Press, 1977); Spencer R. Weart, *Scientists in Power* (Cambridge: Harvard University Press, 1979); Harry Collins and Trevor Pinch, *The Golem: What You Should Know about Science, Second Edition* (New York: Cambridge University Press, 1993, 1998); Harry Collins and Trevor Pinch, *The Golem at Large: What You Should Know About Technology* (New York: Cambridge University Press, 1998); Wiebe E. Bijker, *Of Bicycles, Bakelites, and Bulbs: Toward a Theory of Sociotechnical Change* (Cambridge: The MIT Press, 1995); Arnold Pacey, *Meaning in Technology* (Cambridge: MIT Press, 1999); Kenneth Lipartito, "Picturephone and the Information Age: The Social Meaning of Failure," *Technology and Culture* 44, no. 1 (Jan. 2003); Charles Perrow, *Normal Accidents: Living with High-Risk Technologies* (New York: Basic Books, Inc., 1984).

26. For a few examples of envirotech literature, see Martin Reuss and Stephen H. Cutcliffe, eds., *The Illusory Boundary: Environment and Technology in History*

(Charlottesville: University of Virginia Press, 2010); Richard White, *The Organic Machine: The Remaking of the Columbia River* (New York: Hill and Wang, 1995); Edmund Russell, "The Garden in the Machine: Toward an Evolutionary History of Technology," in *Industrializing Organisms: Introducing Evolutionary History*, ed. Susan R. Schrepfer and Philip Scranton (New York: Routledge, 2004); Dolly Jørgensen, Finn Arne Jørgensen, and Sara B. Pritchard, eds., *New Natures: Joining Environmental History with Science and Technology Studies* (Pittsburgh: University of Pittsburgh Press, 2013).

27. Richard G. Hewlett and Oscar E. Anderson Jr., *The New World, 1939/1946: Volume I of a History of the United States Atomic Energy Commission* (University Park: The Pennsylvania University Press, 1962), ix. In contrast, Jacques Ellul claimed that atomic energy, atomic weapons in particular, were neither good nor evil but instead "necessary" due to the fact that the technology existed. Jacques Ellul, *The Technological Society*, trans. John Wilkinson (New York: Alfred A Knopf, 1964), 98–99.

CHAPTER ONE

1. Harry S. Truman Presidential Library, Independence, Missouri (hereafter HSTL), Papers of Truman, President's Secretary's Files, Box 174, Folder Atomic Bomb, Press Releases [1 of 3], War Department Press Release on New Mexico Test Site, 3.

2. Historian Ferenc Szasz even went as far to say, "If one examines the list of desired ideal weather conditions gathered from the group leaders and compares them with the actual conditions at the time of the July 16 shot, the contrast is striking." Ferenc M. Szasz, *The Day the Sun Rose Twice: The Story of the Trinity Nuclear Explosion, July 16, 1945* (Albuquerque: University of New Mexico Press, 1985), 77–78.

3. HSTL, Papers of Truman, President's Secretary's Files, Box 174, Folder Atomic Bomb, Press Releases [1 of 3], War Department Press Release on New Mexico Test Site, 1–3.

4. Szasz, *The Day the Sun Rose Twice*, 5.

5. Szasz, *The Day the Sun Rose Twice*, 67–74.

6. Simo Laakkonen, Viktor Pál, and Richard Tucker, "The Cold War and Environmental History: Complementary Fields," *Cold War History* 16, no. 4 (Fall 2016): 394. The article is the introduction to a special issue in the journal focusing on the topic of militarized landscapes.

7. This chapter largely does not consider the human elements of testing, but many historians have been rightly critical of nuclear weapons and the people who tested them. Howard Ball, *Justice Downwind: America's Atomic Testing Program in the 1950s* (New York: Oxford University Press, 1986); Phillip L. Fradkin, *Fallout: An American Nuclear Tragedy* (Tucson: University of Arizona Press, 1989); Richard L. Miller, *Under the Cloud: The Decades of Nuclear Testing* (New York: Free Press, 1986); Barton Hacker generally took a kinder view, in Barton C. Hacker, *Elements of Controversy: The Atomic Energy Commission and Radiation Safety in Nuclear Weapons Testing, 1947–1974* (Berkeley: University of California Press, 1994), 9. Also see Debra Rosenthal, *At the Heart of*

the Bomb: The Dangerous Allure of Weapons Work (Reading: Addison-Wesley Publishing, Inc., 1990); *Nuclear Wastelands: A Global Guide to Nuclear Weapons Production and Its Health and Environmental Effects*, Arjun Makhijani, Howard Hu, and Katherine Yih, eds. (Cambridge: MIT Press, 1995).

8. On real effects, see John Hersey, *Hiroshima* (New York: Bantam Books, Inc., 1946, 1981); Ralph H. Lutts, "Chemical Fallout: Rachel Carson's *Silent Spring*, Radioactive Fallout, and the Environmental Movement," *Environmental Review: ER*, 9, no. 3 (Autumn 1985), 210–25. On imagined effects such as the Hollywood movie *Them!* (1954); see: William M. Tsutsui, "Looking Straight at *Them!* Understanding the Big Bug Movies of the 1950s," *Environmental History* 12, no. 2 (April 2007).

9. Barry Commoner, "The Fallout Problem," *Science* 127, no. 3305 (May 2, 1958): 1023–26.

10. Michael Egan, *Barry Commoner and the Science of Survival: The Remaking of American Environmentalism* (Boston: The MIT Press, 2007), 3.

11. On the decision to drop nuclear bombs on Japan, see J. Samuel Walker, *Prompt and Utter Destruction: Truman and the Use of Atomic Bombs Against Japan* (Chapel Hill: University of North Carolina Press, 1997, 2004), 5, 92–96; J. Samuel Walker, ed., *Nuclear Energy and the Legacy of Harry S. Truman* (Kirksville: Truman State University Press, 2016).

12. HSTL, Papers of George M. Elsey, Box 113, Telegram from Secretary of War to President, 6 August 1945.

13. Some of the earliest appraisals of the atomic bombs dropped on Hiroshima and Nagasaki can be found here in the Strategic Bombing Survey. Most striking are the photographs of destruction wrought. HSTL, White House Central Files, Confidential Files, Box 4, Folder Atomic bomb and energy aug 45 to nov 49, 1 of 2, The United States Strategic Bombing Survey: The Effects of Atomic Bombs on Hiroshima and Nagasaki, 30 June 1946.

14. The National Archives at College Park, Maryland (hereafter NACP), RG 326 Records of the Atomic Energy Commission, Entry A1 19, Minutes of the Meetings of the AEC, Box 1, Meeting no. 100, 24 September 1947, 220.

15. *Fifth Semiannual Report of the Atomic Energy Commission, January 1949*, 5.

16. HSTL, White House Central Files, Confidential Files, Box 4, Folder Atomic bomb and energy aug 45 to nov 49, 2 of 2, Letter from Acting Secretary of the Interior to President, 10 August 1945, 1.

17. HSTL, Papers of Truman, President's Secretary's Files, Box 193, Folder Atomic Bomb 1945, Executive Order 9613, Withdrawing and Reserving for the Use of the United States Lands Containing Radio-Active Mineral Substances, 13 September 1945. A later executive order, 9908, added a clause to all land sales conducted by the United States that reserved for the United States the mineral rights for all radioactive ores contained in such lands. HSTL, Papers of Truman, President's Secretary's Files, Box 194, Folder Atomic Energy, Executive Order 9908, 8 December 1947.

18. To emphasize the early and ever-increasing extent of mining for radioactive ores during the studied period, two examples suffice: (1) As of mid-1952, workers had laid 783 miles of roads for the purpose of accessing radioactive ore deposits at a cost of $4,200,000 (around $5,360 per mile). *Twelfth Semiannual Report of the Atomic Energy Commission, July 1952*, 3. (2) When Eisenhower entered office, US uranium procurement stood at 2,900 total tons each year (990 from the United States, 225 from Canada, and 1,685 from overseas). When Ike left office, that tonnage stood at 33,500 per year (17,730 from the United States, 11,310 from Canada, and 4,460 from overseas). *Annual Report to Congress of the Atomic Energy Commission for 1959, January 1960*, 56. *Annual Report to the Congress of the Atomic Energy Commission for 1960, January 1961*, 113.

19. Gabrielle Hecht, *Being Nuclear: Africans and the Global Uranium Trade* (MIT Press, 2012), 49–50.

20. On the effects of such mining on rural communities in the Colorado Plateau, see: Stephanie A. Malin, *The Price of Nuclear Power: Uranium Communities and Environmental Justice* (New Brunswick: Rutgers University Press, 2015). On how "The [U.S.] Interior Department oversaw an ever-widening quest for minerals that, far from adhering to standard political borders, began in indigenous lands of the American West, circled the Global South, plumbed the oceans, and eventually departed the atmosphere with the leap into outer space." see Megan Black, *The Global Interior: Mineral Frontiers and American Power* (Harvard University Press, 2018), 2.

21. HSTL, Papers of Harry S. Truman, Official File, Box 1527, Folder 692-A MISCELLANEOUS. (Apr.-Oct. 1945), Letter Schuyler Otis Bland to Truman, 24 November 1945.

22. HSTL, Papers of Harry S. Truman, Official File, Box 1527, Folder 692-A MISCELLANEOUS (Apr.-Oct. 1945), Letter Truman to Bland, 26 November 1945.

23. For examples of other protests, see a newspaper article that claimed "the atomic volcano over which the world lives is rumbling and smouldering." HSTL, Papers of Clark Clifford, Subject File, 1945–54, Box 1, Folder Atomic Energy—Newspaper Clippings and Releases, "Smoke from Vesuvius," *Dayton News*, 20 April 1948; worries over harm to fish or the creation of a tidal wave from HSTL, Papers of Harry S. Truman, Official File, Box 1528, 692-A Miscellaneous (Jan.–Apr. 1946); letter Mrs. M. Conan to Truman, 3 April 1946; and many letters to the president, including one from Congressman Vaughn Gary (D-VA), about how the tests might harm the surrounding biota, all in HSTL, Papers of Harry S. Truman, Official File, Box 1528, Folder 692-A Miscellaneous (May–Dec. 1946); seeletter Vaughn Gary to President, 17 June 1946; letter R. Maxwell Bradner to President, 2 July 1946; letter Rolf Kreitz to President, 2 July 1946; letter Walter G. Gleassen to President, 9 July 1946. Several of these letters used explicitly religious reasoning for being against the use of animals during testing. Similar protestations against animals used during nuclear testing also occurred during the Eisenhower presidency. In particular, an article in the *Denver Post* about the AEC Division

of Biology and Medicine using dogs for experiments elicited a great many telegrams from Colorado. Dwight D. Eisenhower Presidential Library, Abilene, Kansas (hereafter DDEL), White House Central Files, General File, Box 1213, Folder 155 1955, Various Telegrams, 26 April 1955. One of the more amusing stories to come out of these tests centers around a journalist who, seeing the tests at Bikini that summer, asked if he had the security clearance to write about how the mushroom cloud produced at Alamogordo had been purple. After an argument ensued about the actual color of the cloud (it had been white), one of the security clearance personnel reminded the journalist that he had been wearing purple sunglasses the day of the test at Alamogordo. HSTL, Dean G. Acheson Papers, Box 2, Folder Atomic Energy, 1947–1948, Control of Atomic Energy, Address by H. Thomas Austern at NYU on 20 March 1948, Washington Square College Alumni, 1.

24. Newsreel footage and a video of the blasts can be found on the Internet Archive. Department of Navy, "Operation Crossroads," https://archive.org/details/gov. ntis.ava13712vnb1, accessed on 15 June 2017.

25. HSTL, Papers of Harry S. Truman, Official File, Box 1533, Folder 692-F The President's Committee to Observe the Atomic Bomb Tests, Press Release of Preliminary Report on 1 July 1946 Bikini Atoll tests, 11 July 1946. The animals that many people protested proved paramount in determining the radiation damage that would have been suffered by ship crews, as "measurements of radiation intensity and a study of animals exposed in ships show that the initial flash of principal lethal radiations, which are gamma-rays and neutrons, would have killed almost all personnel normally stationed aboard the ships centered around the air burst and many others at greater distances." HSTL, Papers of Harry S. Truman, Official File, Box 1527, Folder OF 692-A Atomic Bomb, Preliminary Report Following the Second Atomic Bomb Test, 30 July 1946, 1.

26. HSTL, Papers of Harry S. Truman, Official File, Box 1527, Folder OF 692-A Atomic Bomb, Preliminary Report Following the Second Atomic Bomb Test, 30 July 1946, 2–3.

27. HSTL, Papers of Harry S. Truman, Official File, Box 1533, Folder 692-F The President's Committee to Observe the Atomic Bomb Tests, Report Carl Hatch to President, 29 July 1946, 1–3. On the Crossroads tests, also see HSTL, Papers of Harry S. Truman, Official File, Box 1533, Folder 692-G Joint Chiefs of Staff Evaluation Board, Preliminary Report Following the Second Atomic Bomb Test.

28. On "the hit," historian Michael Sherry has written about "the psychic ease with which men could wage war by air." Quoting Will Irwin's "The Next War": An Appeal to Common Sense (1921), he writes, "The 'gallant' airmen [Irwin] talked to during World War I 'were thinking and talking not of the effects of their bombs but only of 'the hit.'" Michael Sherry, The Rise of American Air Power: The Creation of Armageddon (New Haven: Yale University Press, 1987), 32.

29. HSTL, Papers of Truman, President's Secretary's Files, Box 176, Folder Atomic Testing, Crossroads, The Evaluation of the Atomic Bomb as a Military Weapon: The

Final Report of the Joint Chiefs of Staff Evaluation Board for Operation Crossroads, 30 June 1947, 16, 19–21.

30. Pamela M. Henson, "The Smithsonian Goes to War: The Increase and Diffusion of Knowledge in the Pacific," in *Science and the Pacific War: Science and Survival in the Pacific, 1939–1945*, ed. Roy M. MacLeod (Hingham: Kluwer Academic Publishers, 2000), 41–45. Also see Laura J. Martin, "Proving Grounds: Ecological Fieldwork in the Pacific and the Materialization of Ecosystems," *Environmental History* 23, no. 3 (Jul. 2018): 567–92.

31. HSTL, White House Central Files, Confidential Files, Box 4, Folder Atomic bomb and energy aug 45 to nov 49, 2 of 2, memo from D. E. Lilienthal to President, 18 July 1947; HSTL, White House Central Files, Confidential Files, Box 60, Folder Atomic Energy Commission, Cross Reference Sheet, re "establishing a Proving Ground in the Pacific for routine experiments and tests of atomic weapons," 19 July 1947.

32. John Bordsen, "NC 12 is secret to the Outer Banks Popularity," *Raleigh (NC) News and Observer*, 6 May 2016.

33. HSTL, Papers of Truman, President's Secretary's Files, Box 176, Folder Atomic Testing, Crossroads, Proposed Plan for Atomic Bomb Test Against Naval Targets.

34. NACP, RG 326 Records of the Atomic Energy Commission, Entry A1 19, Minutes of the Meetings of the AEC, Box 2, Meeting no. 193, 16 September 1948, 16.

35. Other than the quotation about a radioactive cloud being used to determine an atomic bomb's composition, all citations in this paragraph from HSTL, Papers of Clark Clifford, Subject File, 1945–54, Box 1, Folder Atomic Energy—Newspaper clippings and Releases, Press Release on the Establishment of Pacific Experimental Installations, 1 December 1947.

36. Emery Jerry Jessee, "Radiation Ecologies: Bombs, Bodies, and Environment during the Atmospheric Nuclear Weapons Testing Period, 1942–1965," PhD diss., Montana State University, 2013, 140–43.

37. Robert E. Kohler, *Landscapes and Labscapes: Exploring the Lab-Field Border in Biology* (Chicago: University of Chicago Press, 2002), xiv, 1. A significant literature exists on the literature about field science as an outdoor laboratory. A few of the most relevant works, in addition to Kohler, include Scott Kirsch, "Ecologists and the Experimental Landscape: The Nature of Science at the US Department of Energy's Savannah River Site," *Cultural Geographies* 14, no. 4 (October 2007): 485–510; Jens Lachmund, "Exploring the City of Rubble: Botanical Fieldwork in Bombed Cities in Germany after World War II," *Osiris* 18 (2003): 234–54; Megan Raby, *American Tropics: The Caribbean Roots of Biodiversity Science* (Chapel Hill: University of North Carolina Press, 2017); Jeremy Vetter, *Field Life: Science in the American West during the Railroad Era* (Pittsburgh: University Pittsburgh Press, 2016).

38. *Eighth Semiannual Report of the Atomic Energy Commission, July 1950*, 119–22.

39. For a more comprehensive history of testing on Amchitka, see: Dean W. Kolhoff,

Amchitka and the Bomb: Nuclear Testing in Alaska (Seattle: University of Washington Press, 2002).

40. HSTL, Papers of Truman, President's Secretary's Files, Box 176, Folder Atomic Energy, Superbomb Data, Memo from James S. Lay, Jr. to Secretary of State, Secretary of Defense, and Chairman of the Atomic Energy Commission, Subj: Underground and Surface Atomic Bomb Tests, 30 October 1950.

41. HSTL, Papers of Truman, President's Secretary's Files, Box 176, Folder Atomic Energy, Superbomb Data, Memo from James S. Lay, Jr. to President, 27 October 1950, 1–2.

42. HSTL, Papers of Truman, President's Secretary's Files, Box 176, Folder Atomic Energy, Superbomb Data, Memo from James S. Lay, Jr. to President, 27 October 1950, 3.

43. HSTL, Papers of Truman, President's Secretary's Files, Box 176, Folder Atomic Energy, Superbomb Data, Memo from James S. Lay, Jr. to President, 27 October 1950, 4.

44. HSTL, Papers of Truman, President's Secretary's Files, Box 176, Folder Atomic Energy, Superbomb Data, Letter from Dale E. Doty to James S. Lay, Jr., 13 October 1950, 1.

45. HSTL, Papers of Truman, President's Secretary's Files, Box 176, Folder Atomic Energy, Superbomb Data, Letter from Dale E. Doty to James S. Lay, Jr., 13 October 1950, 2–3.

46. HSTL, Papers of Truman, President's Secretary's Files, Box 176, Folder Atomic Energy, Superbomb Data, Memo from James S. Lay, Jr. to President, 27 October 1950, 7.

47. HSTL, Papers of Truman, President's Secretary's Files, Box 176, Folder Atomic Testing, Windstorm, Memo from G. C. Marshall to Executive Secretary, National Security Council, Subj: Operation WINDSTORM, 21 May 1951.

48. *Tenth Semiannual Report of the Atomic Energy Commission, July 1951*, 3.

49. HSTL, Papers of Truman, President's Secretary's Files, Box 176, Folder Atomic Energy, Superbomb Data, Letter Morse Salisbury to Joseph Short, Subj: Draft Press Release on New Tests, 19 March 1952, 2. Morse was the Director of the Division of Information Services of the AEC and Short was the Press Secretary to the White House.

50. Kristine C. Harper, "Research from the Boundary Layer: Civilian Leadership, Military Funding and the Development of Numerical Weather Prediction (1946–55)," *Social Studies of Science* 33, no. 5, Earth Sciences in the Cold War (Oct. 2003), 667.

51. *Tenth Semiannual Report of the Atomic Energy Commission, July 1951*, 6.

52. HSTL, Papers of Truman, President's Secretary's Files, Box 176, Folder Atomic Energy, Superbomb Data, Letter Morse Salisbury to Joseph Short, Subj: Draft Press Release on New Tests, 19 March 1952, 3–4.

53. *Thirteenth Semiannual Report of the Atomic Energy Commission, January 1953*, 77–97. And it is an important reminder that these atomic bomb tests fundamentally were about producing data for scientists to analyze. Take, for example, the massive data sets on subjects like subsurface temperature of the bomb range in Las Vegas, Nevada (June 1951), or seismological readings at Amchitka. See The National Archives at

College Park, MD (NACP), RG 374 Records of the Defense Threat Reduction Agency, Box 6, Folder Original Field Notes, Proj 1(8)a, Opn B-J or NACP, RG 374 Records of the Defense Threat Reduction Agency, Box 6, Folder Records from the Seismic Survey of Amchitka Island. Reference Program 8.1 of Windstorm Operation. Classified Information.

54. For past conceptions of human weather control, see James Rodger Fleming, *Fixing the Sky: The Checkered History of Weather and Climate Control* (New York: Columbia University Press, 2010).

55. NACP, RG 326 Records of the Atomic Energy Commission, Entry A1 19, Minutes of the Meetings of the AEC, Box 6, Meeting No. 869, 26 May 1953, 325.

56. The letter's author also asked if Eisenhower could send his farmer friend a letter, as Eisenhower was the first president in forty-five years not to respond to the farmer's letters. This would be "Just a gesture in the spirit of Democracy." DDEL, White House Central Files, General File, Box 1213, Folder 155 Atomic Energy, 1952–53, Letter Bertram G. Frazier to Sherman Adams, 2 June 1953.

57. DDEL, White House Central Files, General File, Box 1214, Folder 155-A, Atomic Bomb, Letter George R. Carr to James C. Haggerty, 9 June 1953.

58. Though this book does not cite anything specifically, see *Fourteenth Semiannual Report of the Atomic Energy Commission, July 1953*, 53. One of the more intriguing resources for studying the 1,350 square mile (860,000 acres) Nevada Proving Grounds is *The Nevada Test Site: A Guide to America's Nuclear Proving Ground* (Culver City, CA: The Center for Land Use Interpretation, 1996).

59. NACP, RG 326 Records of the Atomic Energy Commission, Entry A1 19, Minutes of the Meetings of the AEC, Box 6, Meeting No. 875, 10 June 1953, 357–58.

60. NACP, RG 326 Records of the Atomic Energy Commission, Entry A1 19, Minutes of the Meetings of the AEC, Box 6, Meeting No. 877, 17 June 1953, 365–66.

61. NACP, 326 Records of the Atomic Energy Commission, Entry 73, Division of Biology and Medicine, Records Relating to Fallout Studies, 1953–6, Box 2, Folder Conference—Effect of Atomic Weapons on Weather.

62. Eisenhower even may have been willing to use nuclear weapons during the Korean War. Michael Gordon Jackson, "Beyond Brinkmanship: Eisenhower, Nuclear War Fighting, and Korea, 1953–1968," *Presidential Studies Quarterly* 35, no. 1 (Mar. 2005), 52–75.

63. DDEL, White House Central Files, Official File, Box 213, Folder 8, Commencement Address by Thomas E. Murray at Seattle University, Seattle, WA, 29 May 1953, 2. Murray may have been speaking of a general public when he said, "the world was unaware of" uranium that recently. The mineral was discovered long before that.

64. Though a work of fiction, for an understanding of the effects that atomic bombs of ever-increasing power had on the world psyche, see the monologue Alan Moore and Dave Gibbons, *Watchmen* (New York: DC Comics, 1986, 2005), 13.

65. DDEL, Dwight D. Eisenhower Papers as President, Ann Whitman File, Press

Conference Series, Box 2, Folder Press Conference 31 March 1954, Statement by Lewis L. Strauss, 1. Also see *Seventeenth Semiannual Report of the Atomic Energy Commission, January 1955*, 14.

66. Joseph Masco, *The Nuclear Borderlands: The Manhattan Project in Post–Cold War New Mexico* (Princeton University Press, 2006), 294–95.

67. DDEL, Dwight D. Eisenhower Papers as President, Ann Whitman File, Press Conference Series, Box 2, Folder Press Conference 31 March 1954, Official White House Transcript of President Eisenhower's Press and Radio Conference #33, 13.

68. DDEL, Dwight D. Eisenhower Papers as President, Ann Whitman File, Press Conference Series, Box 2, Folder Press Conference 31 March 1954, Statement by Lewis L. Strauss, 2–3.

69. DDEL, Dwight D. Eisenhower Papers as President, Ann Whitman File, Press Conference Series, Box 2, Folder Press Conference 31 March 1954, Statement by Lewis L. Strauss, 2.

70. DDEL, Dwight D. Eisenhower Papers as President, Ann Whitman File, Press Conference Series, Box 2, Folder Press Conference 31 March 1954, Statement by Lewis L. Strauss, 2–4.

71. DDEL, Dwight D. Eisenhower Papers as President, Ann Whitman File, Administrative Series, Box 4, Folder Atomic Energy Commission 1954–54 (2), Letter R. B. Anderson to Eisenhower, 8 December 1954.

72. Valerie Kuletz, *The Tainted Desert: Environmental and Social Ruin in the American West* (New York: Routledge, 1998), 6.

73. NACP, RG 326 Records of the Atomic Energy Commission, Entry A1 19, Minutes of the Meetings of the AEC, Box 8, Meeting No. 1067, 15 March 1955, 166–67.

74. *Eighteenth Semiannual Report of the Atomic Energy Commission*, July 1955, 32.

75. DDEL, Dwight D. Eisenhower Papers as President, Ann Whitman File, Administrative Series, Box 4, Folder Atomic Energy Commission 1955–56 (7), Letter from Lewis L. Strauss to Eisenhower, 28 March 1955.

76. DDEL, Dwight D. Eisenhower Papers as President, Ann Whitman File, Press Conference Series, Box 4, Press Conference 25 April 1956, Official White House Transcript of President Eisenhower's Press and Radio Conference #85, 8.

77. *Twentieth Semiannual Report of the Atomic Energy Commission, July 1956*, 8, 115. See also *Twenty-First Semiannual Report of the Atomic Energy Commission, January 1957*, 9–11.

78. DDEL, White House Central Files, General File, Box 1215, Folder 155-B, Sept. to date 1956 (1), Letter Dorothy Leslie to the President, 21 October 1956 and Letter Howard Pyle to Dorothy Leslie, 30 October 1956.

79. DDEL, White House Central Files, General File, Box 1215, Folder 155-B, Sept. to date 1956 (1), Letter Mrs. Mitchell Fine to President, 26 October 1956 and Letter Sherman Adams to Mrs. Mitchell Fine, 31 October 1956. This letter from Lewis Strauss to

President Eisenhower also stressed the need for testing to improve weapons for the "defense stockpile" and reducing fallout or radiation "contamination which might result from the spread of nuclear material from weapons involved in fire or accident." DDEL, Dwight D. Eisenhower Papers as President, Ann Whitman File, Administrative Series, Box 4, Folder Atomic Energy Commission 1955–56 (1), Letter Lewis Strauss to Eisenhower, 21 December 1956.

80. DDEL, White House Central Files, General File, Box 1215, Folder 155-B, Jan.—Aug. 1957 (2), Letter R. M. Tildesley to President, 29 June 1957. For other examples, see DDEL, White House Central Files, General File, Box 1215, Folder 155-B, Jan.–Aug. 1957 (2), Petition from University of Washington students, 24 June 1957 and DDEL, White House Central Files, General File, Box 1215, Folder 155-B, Jan.–Aug. 1957 (2), Letter R. M. Tildesley to President, 29 June 1957.

81. DDEL, Dwight D. Eisenhower Papers as President, Ann Whitman File, Administrative Series, Box 4, Folder Atomic Energy Commission 1955–56 (2), Statement by the President, 24 October 1956.

82. Other projects also had a distinctly environmental or biological component. DDEL, John Stewart Bragdon Records, Miscellaneous File, Box 1, Folder Atomic Energy Commission, Operation Plumbbob Civil Effects Test Group Project Summaries, 1957, ii, 42, 43, 47. On the animal testing in Operation Plumbbob, where especially pigs but also dogs, sheep, cows, monkeys, and mice were used, see Masco. As he writes, "The protected body of the Cold Warrior, increasingly rendered as cyborg in the cockpit of plans and other military machines, was thus prefigured by the vaporized, mutilated, and traumatized animal body." Masco, *The Nuclear Borderlands*, 307–11.

83. NACP, RG 326 Records of the Atomic Energy Commission, Entry A1 19, Minutes of the Meetings of the AEC, Box 10, Meeting No. 1277, 17 April 1957, 193–95.

84. *Twenty-fourth Semiannual Report of the Atomic Energy Commission, July 1958*, 14.

85. *Twenty-third Semiannual Report of the Atomic Energy Commission, January 1958*, 274–75.

86. The "Suit to Enjoin Nuclear Tests," filed by a group of fourteen people from five different countries, used specifically environmental reasoning. DDEL, White House Central Files, Official File, Box 451, Folder OF 108-A Atomic Weapons, Atomic and Hydrogen Bombs (9), Letter from George Cochran Doub to H. Roemer McPhee, 23 April 1958. The hunger strike at AEC headquarters occurred in May 1958. NACP, RG 326 Records of the Atomic Energy Commission, Entry A1 19, Minutes of the Meetings of the AEC, Box 11, Meeting No. 1372, 9 May 1958. And the citizen worried about the moon, James H. Ouzts, was quite serious, even though it appears comical now. DDEL, White House Central Files, General File, Box 1216, Folder 155-B, Apr.–June 1959, Telegram James H. Ouzts to President, 23 March 1959.

87. *Twenty-fourth Semiannual Report of the Atomic Energy Commission, July 1958*, 13.

88. DDEL, Dwight D. Eisenhower Papers as President, Ann Whitman File, Press

Conference Series, Box 8, Press Conference 17 June 1959, Official White House Transcript of President Eisenhower's Press and Radio Conference, #161, 4–5.

89. DDEL, White House Office, Office of the Special Assistant, OCB Series, Subject Subseries, Box 5, Folder Nuclear Energy Matters (5) [Apr–Oct 1958], Pacific Nuclear Tests Concluded, 3 September 1958.

90. The pretest estimations claimed, "The detonations are not expected to add enough radioactive material to natural levels of radioactivity in the ocean to be harmful to marine life." Either way, testing of the waters and marine life occurred, as well as studies into the "ultimate destination and behavior of radioactivity in the sea water and in marine organisms." *Twenty-fourth Semiannual Report of the Atomic Energy Commission, July 1958*, 348–50.

91. Robert A. Divine, "Eisenhower, Dulles, and the Nuclear Test Ban Issue: Memorandum of a White House Conference, 24 March 1958." *Diplomatic History* 2, no. 1 (October 1978), 327.

92. The names of tests in the Hardtack I test series, conducted from 28 April to 18 August 1958, were, in order of testing date Yucca, Cactus, Fir, Butternut, Koa, Wahoo, Holly, Nutmeg, Yellowwood, Magnolia, Tobacco, Sycamore, Rose, Umbrella, Maple, Aspen, Walnut, Linden, Redwood, Elder, Oak, Hickory, Sequoia, Cedar, Dogwood, Poplar, Scaevola, Pisonia, Juniper, Olive, Pine, Teak, Quince, Orange, Fig.

CHAPTER TWO

1. Scholars have analyzed the film many times. See, for example, chapter 1, "Government Propaganda Films and Civil Defense," in Melvin E. Matthews Jr., *Duck and Cover: Civil Defense Images in Film and Television from the Cold War to 9/11* (Jefferson: McFarland & Company, Inc., 2012), 11–33.

2. "Duck and Cover," directed by Anthony Rizzo (1952; Archer Productions). http://www.youtube.com. Accessed 17 April 2012.

3. President Eisenhower reminded the nation in 1956 that the US government knew about radiation since Trinity. Dwight D. Eisenhower Presidential Library, Abilene, Kansas (hereafter DDEL), Dwight D. Eisenhower Papers as President, Ann Whitman File, Administrative Series, Box 4, Folder Atomic Energy Commission 1955–56 (2), Statement by the President, 24 October 1956. And an October 1946 memo from Colonel Stafford Warren (in charge of radiological safety at Operation Crossroads) to General Lesley Groves (head of the Manhattan Project) stated that radioactive fallout can and will cause cancer if there is significant exposure. Howard Ball, *Justice Downwind: America's Atomic Testing Program in the 1950s* (New York: Oxford University Press, 1986), 204. On Julian H. Webb's discovery, see Matt Blitz, "When Kodak Accidentally Discovered A-Bomb Testing," *Popular Mechanics*, 20 June 2016, http://www.popularmechanics.com. His 1949 report on tainted film explicitly connected the problems Kodak experienced with film "fogging" to the 1945 Trinity test. Eventually, after threats of a lawsuit, the AEC provided Kodak with future test schedules and locations so that the

company could take any necessary precautions to safeguard its products. For more on Kodak and other tests, see E. Jerry Jessee, "A Heightened Controversy: Nuclear Weapons Testing, Radioactive Tracers, and the Dynamic Stratosphere," in *Toxic Airs: Body, Place, Planet in Historical Perspective*, James Rodger Fleming and Ann Johnson, eds. (University of Pittsburgh Press, 2014), 152–53.

4. While both common citizens and government officials may have known about fallout radiation much earlier, many of the most important historical works on the subject did not appear until the 1980s. On images of radioactive fallout in US culture, see chapter 11, "Death Dust," in Spencer R. Weart, *The Rise of Nuclear Fear* (Cambridge: Harvard University Press, 2012). Some other useful historical works on fallout include Ralph H. Lutts, "Chemical Fallout: Rachel Carson's Silent Spring, Radioactive Fallout, and the Environmental Movement," *Environmental Review: ER* 9, no. 3 (Autumn 1985), 210–25; Richard L. Miller, *Under the Cloud: The Decades of Nuclear Testing* (New York: The Free Press, 1986); Paul Boyer, *Fallout: A Historian Reflects on America's Half-Century Encounter with Nuclear Weapons* (Columbus: Ohio State University Press, 1998); Robert A. Jacobs, *The Dragon's Tail: American's Face the Atomic Age* (Amherst and Boston: University of Massachusetts Press, 2010); Ball, *Justice Downwind*; Philip L. Fradkin, *Fallout: An American Nuclear Tragedy* (Tucson: University of Arizona Press, 1989); Barton C. Hacker, *Elements of Controversy: The Atomic Energy Commission and Radiation Safety in Nuclear Weapons Testing, 1947–1974* (Berkeley: University of California Press, 1994).

5. Chapter 6, "Ecology and the Atomic Age," in Joel B. Hagen, *An Entangled Bank: The Origins of Ecosystem Ecology* (New Brunswick: Rutgers University Press, 1992), 100–121.

6. He argued, "One result of this attitude is that even precautions that could have been taken often were not. No one warned people living downwind from Hanford not to drink locally produced milk, which was tainted with iodine-131. Downwind from the Nevada Test Site, the AEC routinely reassured people that there was little or no danger, despite knowledge of heavy fallout after many tests." Arjun Makhijani, Howard Hu, and Katherine Yih, eds. *Nuclear Wastelands: A Global Guide to Nuclear Weapons Production and Its Health and Environmental Effects*, (Cambridge: MIT Press, 1995), 1, 6.

7. Keith Andrew Meyers, "Investigating the Economic Consequences of Atmospheric Nuclear Testing" (PhD diss. University of Arizona, 2018), 23.

8. Harry S. Truman Presidential Library, Independence, Missouri (hereafter HSTL), Atomic Bomb Collection, Box 1, Folder Franklin D. Roosevelt Library—Copies of Correspondence Between Franklin D. Roosevelt and Albert Einstein [GHDC272], Letter from Albert Einstein to F.D. Roosevelt, 2 August 1939. Dr. Alexander Sachs actually delivered the letter to the White House on 11 October 1939, leading to FDR's response a week later. HSTL, Atomic Bomb Collection, Box 1, Folder Franklin D. Roosevelt Library—Copies of Correspondence Between Franklin D. Roosevelt and Albert Einstein [GHDC272], Letter from Franklin D. Roosevelt to Albert Einstein, 19 October 1939.

9. All citations from the previous paragraph, other than the Einstein-Roosevelt

correspondence, can be found here: Richard G. Hewlett and Oscar E. Anderson, Jr., *The New World, 1939/1946: Volume I of a History of the United States Atomic Energy Commission* (University Park: Pennsylvania University Press, 1962), 5–7. Full text of the McMahon Bill/Atomic Energy Act of 1946 can be found on pp. 714–22.

10. Richard G. Hewlett and Francis Duncan, *Atomic Shield, 1947/1952: Volume II of A History of the United States Atomic Energy Commission* (University Park: The Pennsylvania State University Press, 1969), xiii, xiv, 672–73.

11. Dwight D. Eisenhower Presidential Library, Abilene, Kansas (hereafter DDEL), Dwight D. Eisenhower Papers as President, Ann Whitman File, Campaign Series, Box 6, Folder Atomic Energy, "General Principles Regarding Atomic Energy Development," 1.

12. *Fifth Semiannual Report of the Atomic Energy Commission, January 1949*, 92–93.

13. *Sixth Semiannual Report of the Atomic Energy Commission, July 1949*, 54. The report also detailed the effects of an atomic bomb blast, such as at Hiroshima, by what happened in concentric circles around the detonation center. On early scientific encounters with radiation, particularly after the discovery of x-rays in 1945 till Trinity, see: Matthew Lavine, *The First Atomic Age: Scientists, Radiations, and the American Public, 1895–1945* (New York: Palgrave Macmillan, 2013).

14. The National Archives at College Park, Maryland (hereafter NACP), RG 374 Records of the Defense Threat Reduction Agency, Box 1, Folder Civil Defense—1951, Letter Herbert Scoville Jr. to L. P. Sharples.

15. Richard L. Miller, *Under the Cloud*, 8–9.

16. NACP, 326 Records of the Atomic Energy Commission, Entry 73, Division of Biology and Medicine, Records Relating to Fallout Studies, 1953–6, Box 14, Folder Fallout—Atmospheric Radioactivity, George T. Anton, "Program of the United States Government in Atmospheric Radioactivity," 7 November 1960, 1–2. On the relationship between the US Weather Bureau and the military, see: Kristine C. Harper, "Research from the Boundary Layer: Civilian Leadership, Military Funding and the Development of Numerical Weather Prediction (1946–55)," *Social Studies of Science* 33, no. 5, Earth Sciences in the Cold War (Oct. 2003), 667–96.

17. Jacob Darwin Hamblin, *Arming Mother Nature: The Birth of Catastrophic Environmentalism* (New York: Oxford University Press, 2013), 102–4.

18. NACP, RG 326 Records of the Atomic Energy Commission, Entry 73 Division of Biology and Medicine: Records Relating to Fallout Studies, 1953–1964, Box 1, Folder Sunshine—Gabriel—General Files, 1951 thru 1953.

19. NACP, 326 Records of the Atomic Energy Commission, Entry 73, Division of Biology and Medicine, Records Relating to Fallout Studies, 1953–6, Box 5, various folders on Project Sunshine Bulletin.

20. NACP, RG 326 Records of the Atomic Energy Commission, Entry A1 19, Minutes of the Meetings of the AEC, Box 6, Meeting No. 865, 21 May 1953, 309–10.

21. NACP, RG 326 Records of the Atomic Energy Commission, Entry A1 19, Minutes of the Meetings of the AEC, Box 6, Meeting No. 865, 21 May 1953, 310–11.

22. Valerie Kuletz, *The Tainted Desert: Environmental and Social Ruin in the American West* (New York: Routledge, 1998), 5–6.

23. NACP, RG 326 Records of the Atomic Energy Commission, Entry A1 19, Minutes of the Meetings of the AEC, Box 6, Meeting No. 875, 10 June 1953, 356–57.

24. For more on the sheep, see Barton C. Hacker, "'Hotter Than a $2 Pistol': Fallout, Sheep, and the Atomic Energy Commission, 1953–1986," in *The Atomic West*, ed. Bruce Hevly and John M. Findlay (Seattle: University of Washington Press, 1998).

25. NACP, RG 326 Records of the Atomic Energy Commission, Entry A1 19, Minutes of the Meetings of the AEC, Box 6, Meeting No. 877, 17 June 1953, 369–70.

26. Philip L. Fradkin, *Fallout: An American Nuclear Tragedy* (Tucson: University of Arizona Press, 1989), 24.

27. It was not until April 1955 that the AEC commissioners discussed the Marshall Islanders' return to their homes, and they found those people in good "general health and morale" and decided they could "be returned to their atoll shortly subject only to a number of limitations on diet and movement." NACP, RG 326 Records of the Atomic Energy Commission, Entry A1 19, Minutes of the Meetings of the AEC, Box 8, Meeting No. 1078, 27 April 1955, 267.

28. "Chapter 1: The Birth of Gojira," in William Tsutsui, *Godzilla on My Mind: Fifty Years of the King of Monsters* (New York: Palgrave Macmillan, 2004).

29. NACP, RG 326 Records of the Atomic Energy Commission, Entry A1 19, Minutes of the Meetings of the AEC, Box 7, Meeting No. 984, 12 May 1954, 167–68.

30. DDEL, White House Central Files, Confidential File, Box 9, Folder Atomic Energy Commission, Letter John C. Bugher to Sherman Adams, 1 June 1954 and Letter John C. Bugher to Jane Nishiwaki, 1 June 1954.

31. DDEL, White House Central Files, Confidential File, Box 9, Folder Atomic Energy Commission, Letter John C. Bugher to Jane Nishiwaki, 1 June 1954.

32. DDEL, White House Central Files, Confidential File, Box 9, Folder Atomic Energy Commission, Letter John C. Bugher to Jane Nishiwaki, 1 June 1954.

33. *Sixteenth Semiannual Report of the Atomic Energy Commission, July 1954*, 51–54.

34. NACP, RG 326 Records of the Atomic Energy Commission, Entry A1 19, Minutes of the Meetings of the AEC, Box 7, Meeting No. 1014, 7 July 1954, 300. Indeed, publicity continued to be a problem in the future as well. On 10 June 1955 an article appeared in the *New York Times* that referred to a Mr. Ralph Lapp as a former AEC official when he in fact never had been connected to the commission. The AEC commissioners declared that they needed to inform people, especially the Joint Committee on Atomic Energy, of this and of Lapp's background "to correct any misapprehension that testimony by him on fall-out would be authoritative." NACP, RG 326 Records of the Atomic Energy Commission, Entry A1 19, Minutes of the Meetings of the AEC, Box 8, Meeting No. 1088, 10 June 1955, 351–352.

35. DDEL, White House Central Files, Official File, Box 450, Folder 10, Correspondence between Dean Rusk and President, Rusk to President, 23 February 1955.

36. DDEL, White House Central Files, Official File, Box 450, Folder 10, Correspondence between Dean Rusk and President, Eisenhower to Rusk, 28 February 1955.

37. DDEL, White House Central Files, Official File, Box 450, Folder 10, Correspondence between Dean Rusk and President, Rusk to President, 11 March 1955.

38. NACP, RG 326 Records of the Atomic Energy Commission, Entry A1 19, Minutes of the Meetings of the AEC, Box 8, Meeting No. 1093, 28 June 1955, 417.

39. *Eighteenth Semiannual Report of the Atomic Energy Commission, July 1955*, 79–81, 90–93. The report also contained information, including a map, on fallout over the United States and how it affected the air and water. For information on how radioactive fallout affected the environment and animals at the Nevada Proving Grounds, see: NACP, 326 Records of the Atomic Energy Commission, Entry 73, Division of Biology and Medicine, Records Relating to Fallout Studies, 1953–6, Box 47, Folder Miscellaneous Reports on NPG and also Folder NPG 1955, Folder NPG 1954, and Box 48, Folder NPG 1953. For more on fish contamination, this time on concerns about radiation contamination of a shipment of yellowfin tuna in Portland, Oregon, see: NACP, 326 Records of the Atomic Energy Commission, Entry 73, Division of Biology and Medicine, Records Relating to Fallout Studies, 1953–6, Box 18, Folder Incident—Japanese Fish—Fallout.

40. NACP, 326 Records of the Atomic Energy Commission, Entry 73, Division of Biology and Medicine, Records Relating to Fallout Studies, 1953–6, Box 47, Folder Miscellaneous Loose Papers Confidential (2 of 2), World-Wide Fallout, 1–4.

41. McNeill borrowed the phrase from Ecclesiastes 1:9–11 in the Christian *Bible*. J. R. McNeill, *Something New Under the Sun: An Environmental History of the Twentieth Century World* (New York: W. W. Norton & Company, 2001).

42. Military relationships with the Weather Bureau extended far beyond nuclear weapons. See: Hamblin, *Arming Mother Nature*, 111–12, 119–20.

43. NACP, 326 Records of the Atomic Energy Commission, Entry 73, Division of Biology and Medicine, Records Relating to Fallout Studies, 1953–6, Box 19, Folder General Files—Meteorology—General.

44. NACP, 326 Records of the Atomic Energy Commission, Entry 73, Division of Biology and Medicine, Records Relating to Fallout Studies, Box 3, 1953–6, Folder Stratospheric Monitoring, Jan. 1958, #2, Constant Level, Balloons as Tracers of Air Motion in Atomic Weapon Tests for the Atomic Energy Commission.

45. NACP, 326 Records of the Atomic Energy Commission, Entry 73, Division of Biology and Medicine, Records Relating to Fallout Studies, 1953–6, Box 2, Folder General Files—Fallout—Collection Analysis, Memorandum from Lester Machta to Hal Hollister, Subj: Oceanic Sampling, 16 November 1959. Machta was Chief, Special Projects Section, Office of Meteorological Research, US Weather Bureau; Hollister was Division of Biology and Medicine of the AEC. Also see October 1960 investigations by the

Scripps Institute into providing "a general description of the distribution of temperature, salinity, density, oxygen, and other chemical properties of the Indian Ocean—in particular the study of the changing circulation pattern under the influence of the monsoon winds." NACP, 326 Records of the Atomic Energy Commission, Entry 73, Division of Biology and Medicine, Records Relating to Fallout Studies, 1953–6, Box 19, Folder Indian Ocean Expedition.

46. NACP, 326 Records of the Atomic Energy Commission, Entry 73, Division of Biology and Medicine, Records Relating to Fallout Studies, 1953–6, Box 17, Folder Ozone contains correspondence on the subject and attaches an October 1960 article in the British journal *Endeavour* called "Radioactive tracers in the atmosphere" by N. G. Stewart.

47. NACP, 326 Records of the Atomic Energy Commission, Entry 73, Division of Biology and Medicine, Records Relating to Fallout Studies, 1953–6, Box 47, Folder Miscellaneous Loose Papers Confidential (2 of 2), World-Wide Fallout, 5–12.

48. For example, in March 1956, the United Nations convened its Scientific Committee on Atomic Radiation, especially concerned with fallout and the "effects of radiation on man and his environment." *Twentieth Semiannual Report of the Atomic Energy Commission, July 1956*, 16. For other various research on fallout, see "Longterm Effects of Fall-Out from Nuclear Weapons" in *Nineteenth Semiannual Report of the Atomic Energy Commission, January 1956*, 69. Also inquiries into dairy products, comparing the United States to the rest of the world in *Twentieth Semiannual Report of the Atomic Energy Commission, July 1956*, 106. Also see work using weather balloons to study Sr90 and other nuclides at altitudes of fifty thousand to ninety thousand feet in NACP, 326 Records of the Atomic Energy Commission, Entry 73, Division of Biology and Medicine, Records Relating to Fallout Studies, 1953–6, Box 3, Folder #1 Stratospheric Monitoring, April 55 through Dec. 57, Letter General Manager to Clinton P. Anderson.

49. NACP, RG 326 Records of the Atomic Energy Commission, Entry A1 19, Minutes of the Meetings of the AEC, Box 9, Meeting No. 1238, 22 October 1956, 638–39.

50. The AEC did admit that some unexpected weather conditions caused heavy radioactive fallout on four Pacific Islands—the Marshall Islands. *Twenty-First Semiannual Report of the Atomic Energy Commission, January 1957*, 112–13.

51. NACP, 326 Records of the Atomic Energy Commission, Entry 73, Division of Biology and Medicine, Records Relating to Fallout Studies, 1953–6, Box 18, Folder Fireball Chemical Project, Memo W. F. Libby to General Manager, 2 April 1957. The memo carbon copied all of the AEC commissioners.

52. NACP, 326 Records of the Atomic Energy Commission, Entry 81 A, Commissioner Harold S. Vance, Correspondence, 1955–1959, Box 2, Folder Memos to Chairman & General Manager, Memo H. S. Vance to Dr. Libby, 6 March 1957.

53. "An Appeal by American Scientists to the Governments and People of the World." DDEL, White House Central Files, Official File, Box 451, Folder OF 108-A Atomic Weapons, Atomic and Hydrogen Bombs (8), Letter Linus Pauling to President, 4 June 1957 and Letter Sherman Adams to Linus Pauling, 29 June 1957. Also

see this petition by University of Washington students. DDEL, White House Central Files, General File, Box 1215, Folder 155-B, Jan.–Aug. 1957 (2), Petition from University of Washington students, 24 June 1957.

54. President Eisenhower later tried to clarify his comments about scientists working outside his field with a fairly conspiracy theory–minded assertion, saying, "I don't know. I haven't any idea, but I just say it seems to come up in so many places and so many different speeches, and you find scientists of various kinds other than geneticists and physicists in this particular field that have something to say about it." All Eisenhower quotations in paragraph from: DDEL, Dwight D. Eisenhower Papers as President, Ann Whitman File, Press Conference Series, Box 6, Press Conference 5 June 1957, Official White House Transcript of President Eisenhower's Press and Radio Conference #112.

55. DDEL, White House Central Files, Official File, Box 451, Folder OF 108-A Atomic Weapons, Atomic and Hydrogen Bombs (7), Letter from Morse Salisbury to James G. Hagerty, 21 June 1957.

56. DDEL, Dwight D. Eisenhower Papers as President, Ann Whitman File, Administrative Series, Box 5, Folder Atomic Energy Commission 1958 (4), "Fallout from Nuclear Weapons Testing," A Summary by Charles L. Dunham, MD, Director, Division of Biology and Medicine, US Atomic Energy Commission.

57. This idea is similar to (and the story continued in the 1980s as well) chapter 2, "Strategic Defense, Phony Facts, and the Creation of the George C. Marshall Institute," in Naomi Oreskes and Erik M. Conway, *Merchants of Doubt: How a Handful of Scientists Obscured the Truth on Issues from Tobacco Smoke to Global Warming* (New York: Bloomsbury Press, 2010).

58. NACP, RG 326 Records of the Atomic Energy Commission, Entry A1 19, Minutes of the Meetings of the AEC, Box 10, Meeting No. 1296, 26 July 1957, 378.

59. *Twenty-second Semiannual Report of the Atomic Energy Commission, July 1957*, Appendix 11 "Remarks Prepared by Dr. Willard F. Libby, Commissioner, For Delivery Before the American Physical Society, Washington, DC, April 26," 223–35.

60. Hamblin, *Arming Mother Nature*, 103.

61. NACP, 326 Records of the Atomic Energy Commission, Entry 73, Division of Biology and Medicine, Records Relating to Fallout Studies, 1953–6, Box 1, Folder Sunshine—General, 1957, Letter Eilif Dahl to W. F. Libby, 18 June 1957. Similarly, a research officer in the USDA sent a letter to a health Physicist in the AEC Division of Biology and Medicine thanking him for translating something from Russian about the behavior of fission products in soil. NACP, 326 Records of the Atomic Energy Commission, Entry 73, Division of Biology and Medicine, Records Relating to Fallout Studies, 1953–6, Box 1, Folder Sunshine—General, 1957, Letter F. D. H. Macdowall to Forrest Western, 14 August 1957.

62. *Twenty-second Semiannual Report of the Atomic Energy Commission, July 1957*, 132–35.

63. Barbara Rose Johnston and Holly M. Barker, *The Rongelap Report: Consequential Damages of Nuclear War* (Walnut Creek: Left Coast Press, 2008), 23.

64. Elizabeth DeLoughrey, "Radiation Ecologies and the Wars of Light," *Modern Fiction Studies* 55, no. 3 (Fall 2009), 469.

65. *Twenty-second Semiannual Report of the Atomic Energy Commission, July 1957*, 132–35.

66. *Twenty-second Semiannual Report of the Atomic Energy Commission, July 1957*, 10–12.

67. NACP, 326 Records of the Atomic Energy Commission, Entry 73, Division of Biology and Medicine, Records Relating to Fallout Studies, 1953–6, Box 1, Folder World-Wide Fallout (Sunshine)—General 1958, Program 37, Civil Effects Test Group, Radio-Ecological Aspects of Nuclear Fallout: Introduction, Methods and Procedures, September 1957. The primary authors of the report were K. H. Larson, J. W. Neel, R. G. Lindberg, L. Baurmash, and G. V. Alexander.

68. Clark, a geologist with the Richfield Oil Corporation in Bakersfield, California, claimed the necessity of studying fallout was truly necessary, though, stating, "Whether the danger is real or imagined, the need of the people for reassurance is a very real and valid objective." DDEL, White House Central Files, Official File, Box 450, Folder 1, Letter from George H. Clark to President, 6 August 1957.

69. DDEL, White House Central Files, General File, Box 1216, Folder 155-B, July–Sept. 1958, Statement on Radioactive Fallout Submitted to the US Atomic Energy Commission by the Advisory Committee on Biology and Medicine, 19 October 1957.

70. *Twenty-third Semiannual Report of the Atomic Energy Commission, January 1958*, 277–89. The report also included as appendix 13 the Advisory Committee on Biology and Medicine's "Statement on Radioactive Fallout," roughly what was discussed above.

71. NACP, 326 Records of the Atomic Energy Commission, Entry 81 A, Commissioner Harold S. Vance, Correspondence, 1955–1959, Box 2, Folder Memos to Chairman & General Manager, Memo from H. S. Vance to Chairman, 10 February 1958.

72. Barry Commoner, "The Fallout Problem," *Science* 127, no. 3305 (May 2, 1958), 1023–1026.

73. DDEL, White House Central Files, General File, Box 1216, Folder 155-B, Apr.–June 1958, Telegram from Alfred N. Phillips to President, 8 May 1958.

74. DDEL, White House Central Files, General File, Box 1216, Folder 155-B, July–Sept. 1958, Letter Worthington to President, 26 August 1958. Worthington self-identified as being from the "Middletown Monthly Meeting of Friends."

75. Fear of fallout from a nuclear war likely caused Secretary of State Dulles to write to Leo A. Hoegh of the Federal Civil Defense Administration asking for any literature on fallout "that could tell Mrs. Dulles, in simple terms, what she should install in the cellar to enable my household to utilize the protection against 'fall-out' which the cellar affords?" Hoegh replied with publications titled "Facts about Fallout Protection" ("states simply the problem and measures to be taken") and also "Home Protection

Exercises." DDEL, John Foster Dulles Papers, General Correspondence and Memoranda Series, Box 5, Folder November 25, 1957–April 3, 1958, Letter Dulles to Leo A. Hoegh, 2 April 1958 and Letter Hoegh to Dulles, 3 April 1958.

76. DDEL, White House Central Files, General File, Box 1216, Folder 155-B, Apr.–June 1958, Remarks by Senator Clifford P. Case, Prepared for Delivery at Luncheon of Advertising Club of NJ Honoring Secretary of Labor James P. Mitchell as New Jersey's Outstanding Citizen of 1957, Essex House, Newark, NJ, Noon, Wed., April 19, 1958.

77. *Twenty-fourth Semiannual Report of the Atomic Energy Commission, July 1958,* 196–201.

78. See research on "Radiobiological Surveys of the Pacific." *Twenty-fifth Semiannual Report of the Atomic Energy Commission, January 1959,* 195.

79. NACP, 326 Records of the Atomic Energy Commission, Entry 73, Division of Biology and Medicine, Records Relating to Fallout Studies, 1953–6, Box 16, Folder Radiation Incidents—South Dakota, 12 February 1959; also Letter G. J. Van Heuvelen to Francis J. Weber, 24 February 1959, with attachment Investigation of Reported Radiation Incident at Belle Fourche, South Dakota. Van Heuvelen was a doctor and State Health Officer for South Dakota, and Weber was Chief, Div. of Radiological Health of the Public Health Service.

80. NACP, 326 Records of the Atomic Energy Commission, Entry 73, Division of Biology and Medicine, Records Relating to Fallout Studies, 1953–6, Box 1, Folder World-Wide Fallout (Sunshine)—General 1958, Draft of Fallout Briefing, 5 March 1959, 1–2.

81. NACP, RG 326 Records of the Atomic Energy Commission, Entry A1 19, Minutes of the Meetings of the AEC, Box 12, Meeting No. 1482, 6 March 1959, 190.

82. DDEL, Dwight D. Eisenhower Papers as President, Ann Whitman File, Cabinet Series, Box 13, Folder Cabinet Meeting of 6 March 1959, Minutes of Cabinet Meeting, 6 March 1959.

83. Jessee, "A Heightened Controversy," 152–53, 155.

84. DDEL, Dwight D. Eisenhower Papers as President, Ann Whitman File, Cabinet Series, Box 13, Folder Cabinet Meeting of 6 March 1959, Minutes of Cabinet Meeting, 6 March 1959.

85. NACP, 326 Records of the Atomic Energy Commission, Entry 73, Division of Biology and Medicine, Records Relating to Fallout Studies, 1953–6, Box 18, Folder General Files—Fallout—Foods, Telegram on strontium in foods.

86. As anthropologist Joseph Masco contended, "Nuclear materials not only disrupt the experience of nation-time (confounding notions of both the present and the future), they also upset the concept of nation-space, in that they demonstrate the permeability, even irrelevance, of national borders to nuclear technologies (to intercontinental missiles and radioactive fallout, for example)." Joseph Masco, *The Nuclear Borderlands: The Manhattan Project in Post–Cold War New Mexico* (Princeton University Press, 2006), 11–12.

87. NACP, RG 326 Records of the Atomic Energy Commission, Entry A1 19,

Minutes of the Meetings of the AEC, Box 12, Meeting No. 1493, 17 April 1959, 255.

88. *Annual Report to Congress of the Atomic Energy Commission for 1959, January 1960*, 191, 239–59, appendix 15, "Fallout from Nuclear Tests at Nevada Test Site," 559–72.

89. *Annual Report to the Congress of the Atomic Energy Commission for 1960, January 1961*, 176–79.

90. NACP, 326 Records of the Atomic Energy Commission, Entry 73, Division of Biology and Medicine, Records Relating to Fallout Studies, 1953–6, Box 14, Folder Fallout—Atmospheric Radioactivity, George T. Anton, "Program of the United States Government in Atmospheric Radioactivity," 7 November 1960, 2–14.

91. For an example of how policymakers sometimes valued national security over human and environmental health, see Kate Brown, *Plutopia: Nuclear Families, Atomic Cities, and the Great Soviet and American Plutonium Disasters* (New York: Oxford University Press, 2012).

CHAPTER THREE

1. Barry Commoner, "Conservation of the Water Resource: The Responsibility of the Scientist," *Journal (Water Pollution Control Federation)* 37, no. 1 (Jan. 1965): 60.

2. As scholar Marek Thee has argued, "The history of the nuclear arms race is intimately interrelated with the systematic testing of nuclear weapons." Marek Thee, "The Pursuit of a Comprehensive Nuclear Test Ban," *Journal of Peace Research* 25, no. 1 (Mar. 1988): 15.

3. Particularly important is Kai-Henrik Barth's research on the important role seismology played in detecting nuclear tests, meaning the discipline was crucial to Eisenhower-era nuclear test cessation treaties. However, it is not merely banal to point out that those detection capabilities took more than just seismology and were particularly based in sweeping knowledge of the natural world. Kai-Henrik Barth, "Detecting the Cold War: Seismology and Nuclear Weapons Testing, 1945–1970" (PhD diss., University of Minnesota, 2000). Studies of the nuclear test ban treaty and its previous negotiations began soon after the inking of the official Limited Test Ban Treaty in 1963. Egon Schwelb, "The Nuclear Test Ban Treaty and International Law," *American Journal of International Law* 58, no. 3 (July 1964): 642–70. Harold K. Jacobson and Eric Stein, *Diplomats, Scientists and Politicians: The United States and the Nuclear Test Ban Negotiations* (Ann Arbor, University of Michigan Press, 1966); James Hubert McBride, *The Test Ban Treaty: Military, Technological, and Political Implications* (Chicago: Henry Regnery Company, 1967), vi; Robert A. Divine, *Blowing on the Wind: The Nuclear Test Ban Debate, 1954–1960* (New York: Oxford University Press, 1978); Per Fredrik Ilsaas Pharo, "A Precondition for Peace: Transparency and the Test-ban Negotiations, 1958–1963," *International History Review* 22, no. 3 (September 2000); Martha Smith-Norris, "The Eisenhower Administration and the Nuclear Test Ban Talks, 1958–1960: Another Challenge to Revisionism," *Diplomatic History* 27, no. 4 (August

2003); Benjamin P. Greene, *Eisenhower, Science Advice, and the Nuclear Test-ban Debate, 1945–1963* (Stanford: Stanford University Press, 2007). While most chronicles have recognized the importance of the Eisenhower administration in the eventual agreement signed by Kennedy, two relative exceptions to this notion are Glenn T. Seaborg, *Kennedy, Khrushchev, and the Test Ban* (Berkeley: University of California Press, 1981) and Kendrick Oliver, *Kennedy, Macmillan and the Nuclear Test-ban Debate, 1961–63* (New York: St. Martin's Press, Inc., 1998); on Eisenhower see also Anthony James Joes, "Eisenhower Revisionism: The Tide Comes In," *Presidential Studies Quarterly* 15, no. 3 (Summer, 1985); Mary S. McAuliffe, "Eisenhower, the President," *Journal of American History* 68, no. 3 (Dec., 1981); Donald Alan Carter, "Eisenhower versus the Generals," *Journal of Military History* 71, no. 4 (Oct., 2007); Elmo Richardson, *The Presidency of Dwight D. Eisenhower* (Lawrence: The Regents Press of Kansas, 1979).

4. Historians Kurk Dorsey and Mark Lytle, in their introduction to a special roundtable issue in *Diplomatic History*, called the separation between environmental and diplomatic histories "increasingly untenable." Kurk Dorsey and Mark Lytle, "Introduction," *Diplomatic History* 32, no. 4 (September 2008), 517. See also Toshihiro Higuchi, "Atmospheric Nuclear Weapons Testing and the Debate on Risk Knowledge in Cold War America, 1945–1963," in *Environmental Histories of the Cold War*, ed. J. R. McNeill and Corinna R. Unger (New York: Cambridge University Press, 2010). And for an investigation of the co-opting of science into Cold War ideological struggles surrounding the test ban treaty debates, see chapter 6, "Science for Diplomacy, in Audra J. Wolfe, *Freedom's Laboratory: The Cold War Struggle for the Soul of Science* (Baltimore: Johns Hopkins University Press, 2018), 113–34.

5. Toshihiro Higuchi, *Political Fallout: Nuclear Weapons Testing and the Making of a Global Environmental Crisis* (Stanford University Press, 2020).

6. On increased global environmental monitoring for military and geopolitical purposes, see chapter 4, "Earth Under Surveillance" in Jacob Darwin Hamblin, *Arming Mother Nature: The Birth of Catastrophic Environmentalism* (New York: Oxford University Press, 2013).

7. One newspaper article described the blast, "In any event the story seems to prove that the destructive potential of the hydrogen bomb has not been exaggerated." "The Unlucky Dragon," *Washington Post and Times Herald*, 19 March 1954, 28.

8. The idea that nuclear energy might be used in peaceful endeavors was not new to Eisenhower. For example, in a late 1952 debate the United Kingdom's Minister of Works, David Eccles, coined the idea of "plutonium into plowshares." DDEL, Papers as President, Ann Whitman File, Speech Series, Box 5, Folder United Nations Speech 8 Dec 1953 (2), Intelligence Report No. 6500, "Official Foreign Reactions to President Eisenhower's Speech of December 8, 1953," 2. Project Plowshare took its name from the *Bible* verse Micah 4:3 (New International Version translation), which reads, "He will judge between many peoples and will settle disputes for strong nations far and wide. They will beat their swords into plowshares and their spears into pruning hooks.

Nation will not take up sword against nation, nor will they train for war anymore." See also Richard G. Hewlett and Jack M. Holl, *Atoms for Peace and War, 1953–1961: Eisenhower and the Atomic Energy Commission* (Berkeley: University of California Press, 1989).

9. DDEL, Papers as President, Ann Whitman File, Speech Series, Box 5, Folder United Nations Speech 8 Dec 1953 (1), "Atomic Power for Peace," An Address by President Eisenhower before The General Assembly of the United Nations, 8 December 1953, 11.

10. DDEL, Dwight D. Eisenhower Papers as President, Ann Whitman File, Administrative Series, Box 4, Folder Atomic Energy Commission 1955–56 (2), Statement by the President, 24 October 1956, 1.

11. DDEL, Dwight D. Eisenhower Papers as President, Ann Whitman File, Administrative Series, Box 4, Folder Atomic Energy Commission 1955–56 (2), Statement by the President, 24 October 1956, 2.

12. DDEL, Dwight D. Eisenhower Papers as President, Ann Whitman File, Administrative Series, Box 4, Folder Atomic Energy Commission 1955–56 (2), Statement by the President, 24 October 1956, 3.

13. DDEL, Dwight D. Eisenhower Papers as President, Ann Whitman File, Administrative Series, Box 4, Folder Atomic Energy Commission 1955–56 (2), Statement by the President, 24 October 1956, 3–4.

14. DDEL, Dwight D. Eisenhower Papers as President, Ann Whitman File, Administrative Series, Box 4, Folder Atomic Energy Commission 1955–56 (2), Statement by the President, 24 October 1956, 4.

15. DDEL, Dwight D. Eisenhower Papers as President, Ann Whitman File, Press Conference Series, Box 6, Press Conference 22 May 1957, prepress conference notes.

16. DDEL, Dwight D. Eisenhower Papers as President, Ann Whitman File, Press Conference Series, Box 6, Press Conference 5 June 1957, Official White House, Transcript of President Eisenhower's Press and Radio Conference #112, 7, 11.

17. Eisenhower again emphasized the importance of "clean" nuclear bombs for peaceful uses in a 3 July 1957 press conference. He said, "We are trying to make small bombs—clean bombs, and to develop usefulness in a peaceful world as well as just weapons of war." DDEL, Dwight D. Eisenhower Papers as President, Ann Whitman File, Press Conference Series, Box 6, Press Conference 3 July 1957, Official White House Transcript of President Eisenhower's Press and Radio Conference #115, 6. Project Plowshare attempted to use nuclear weapons during times of peace as a sort of nuclear dynamite. DDEL, White House Office, Office of the Special Assistant, OCB Series, Subject Subseries, Box 5, Folder Nuclear Energy Matters (5) [Apr–Oct 1958], Memorandum for the Operations Coordinating Board, Subj: Item for Luncheon Meeting, 8 October 1958. For more on the Plowshare program, see Dan O'Neill, *The Firecracker Boys* (New York: St. Martin's Press, 1994); Scott Kirsch, *Proving Grounds: Project Plowshare and the Unrealized Dream of Nuclear Earthmoving* (New Brunswick: Rutgers University Press, 2005).

18. The president further elaborated on Soviet use of clean weapons, "I would hope that they would learn how to use clean bombs and if they ever used any atomic bombs would use clean ones,—for the simple reason that then at least the bombs used would be specific weapons instead of weapons of general and uncontrolled destruction." DDEL, Dwight D. Eisenhower Papers as President, Ann Whitman File, Press Conference Series, Box 6, Press Conference 26 June 1957, Official White House Transcript of President Eisenhower's Press and Radio, Conference, #114, 2–5, 8.

19. The National Archives at College Park, Maryland (hereafter NACP), RG 326 Records of the Atomic Energy Commission, Entry A1 19, Minutes of the Meetings of the AEC, Box 10, Meeting No. 1294, 17 July 1957, 335.

20. Italicized text in this passage appears as underlined text in the original document. DDEL, White House Office, Office of the Special Assistant, OCB Series, Administrative Subseries, Box 2, Folder Operations Coordinating Board General (3) [September–October 1957], Memo from R. Hirsche to Cutler, Subj: AEC Reaction re Hardtack (1958 Pacific Weapons Test) Announcement, 12 September 1957.

21. DDEL, White House Office, Office of the Special Assistant, OCB Series, Subject Subseries, Box 4, Folder Nuclear Energy Matters (2) [Jan 1958], Memo J. H. Morse to F. M. Dearborn, Subject: Reasons for Further Nuclear Testing, 20 January 1958, 1–2.

22. Historian Donald Worster has even argued that the ethos of capitalism, in terms of the environment, has three tenets: 1) "Nature must be seen as capital." 2) "Man has a right, even an obligation to use this capital for constant self-advancement." 3) "The social order should permit and encourage this increase of personal wealth." Donald Worster, *Dust Bowl: The Southern Plains in the 1930s* (New York: Oxford University Press, 1979), 6.

23. DDEL, White House Office, Office of the Special Assistant, OCB Series, Subject Subseries, Box 4, Folder Nuclear Energy Matters (2) [Jan 1958], Memo J. H. Morse to F. M. Dearborn, Subject: Reasons for Further Nuclear Testing, 20 January 1958, 3–6.

24. DDEL, White House Office, Office of the Special Assistant, OCB Series, Subject Subseries, Box 4, Folder Nuclear Energy Matters (2) [Jan 1958], Memo Robert E. Matteson to John Morse, Subject: Test Suspension Issue, 24 January 1958, 1–6.

25. Captain John H. Morse was also sent a copy of the memo on the same day. DDEL, White House Office, Office of the Special Assistant, OCB Series, Subject Subseries, Box 4, Folder Nuclear Energy Matters (4) [Feb-Mar 1958], Memo to President, 14 February 1958.

26. Charles S. Maier, "Introduction: Science, Politics, and Defense in the Eisenhower Era," in *A Scientist at the White House: The Private Diary of President Eisenhower's Special Assistant for Science and Technology*, ed. George B. Kistiakowsky (Cambridge: Harvard University Press, 1976), xxvi.

27. Robert A. Divine, "Eisenhower, Dulles, and the Nuclear Test Ban Issue: Memorandum of a White House Conference, 24 March 1958." *Diplomatic History* 2, no. 1 (October 1978): 321–30.

28. DDEL, White House Office, Office of the Special Assistant, OCB Series, Administrative Subseries, Box 1, Folder Chronological—Karl G. Harr January-March 1958 (4), Memo From R. V. Mrozinski to Staats, Subj: The Psychological Aspects of the forthcoming U.S. Nuclear Tests.

29. Captain Morse went so far as to say that testing should continue underground "Without radioactivity . . . [F]or the benefit of humanity." DDEL, White House Office, Office of the Special Assistant, OCB Series, Subject Subseries, Box 5, Folder Nuclear Energy Matters (5) [Apr–Oct 1958], Memo for Mr. Staats by Manning H. Williams, Subj: Possible OCB Action on Disarmament, 1 April 1958.

30. Remarks by Senator Clifford P. Case, Prepared for Delivery at Luncheon of Advertising Club of NJ Honoring Secretary of Labor James P. Mitchell as New Jersey's Outstanding Citizen of 1957, Essex House, Newark, NJ Noon, Wed., April 9, 1958. Sent to the White house in DDEL, White House Central Files, General File, Box 1216, Folder 155-B, Apr.—June 1958, Letter from Sam Zagoria to James C. Hagerty, 9 April 1958.

31. DDEL, Dwight D. Eisenhower Papers as President, Ann Whitman File, Press Conference Series, Box 7, Press Conference 9 April 1958, prepress conference notes.

32. DDEL, Dwight D. Eisenhower Papers as President, Ann Whitman File, Press Conference Series, Box 7, Press Conference 9 April 1958, Official White House Transcript of President Eisenhower's Press and Radio Conference, #131, 4.

33. Barth, "Detecting the Cold War," 131. The entirety of chapter 5, "'Political Science:' The Geneva Conference" focuses on the Conference of Experts. Also see Divine, Blowing in the Wind, 225–27.

34. Kai-Henrik Barth has argued that the Geneva Conference of Experts was perhaps the crucial moment in the development of US seismology, and that the eventual report was a fundamentally political document. See chapter 5, "'Political Science:' The Geneva Conference of Experts," in Barth, "Detecting the Cold War," 131–61.

35. Eisenhower's 22 August 1958 speech can be heard in its entirety here: http://www.youtube.com. Accessed 15 February 2012.

36. DDEL, White House Central Files, Official File, Cross Reference Sheets, Box 88, Folder OF 108 Atomic Energy 1958, Statement by the President, 25 October 1958.

37. NACP, RG 326 Records of the Atomic Energy Commission, Entry A1 19, Minutes of the Meetings of the AEC, Box 12, Meeting No. 1422, 1 November 1958, 799–800.

38. DDEL, White House Central Files, Official File, Cross Reference Sheets, Box 88, Folder OF 108 Atomic Energy 1958, Statement by the President, 7 November 1958.

39. NACP, RG 326 Records of the Atomic Energy Commission, Entry A1 19, Minutes of the Meetings of the AEC, Box 12, Meeting No. 1432, 24 November 1958, 863–66.

40. NACP, RG 326 Records of the Atomic Energy Commission, Entry A1 19, Minutes of the Meetings of the AEC, Box 12, Meeting No. 1432, 24 November 1958, 866–68.

41. NACP, RG 326 Records of the Atomic Energy Commission, Entry A1 19, Minutes of the Meetings of the AEC, Box 12, Meeting No. 1433, 24 November 1958, 869–71.

42. NACP, RG 326 Records of the Atomic Energy Commission, Entry A1 19, Minutes

of the Meetings of the AEC, Box 12, Meeting No. 1434, 2 December 1958, 872–74.

43. NACP, RG 326 Records of the Atomic Energy Commission, Entry A1 19, Minutes of the Meetings of the AEC, Box 12, Meeting No. 1434, 2 December 1958, 874–75.

44. NACP, RG 326 Records of the Atomic Energy Commission, Entry A1 19, Minutes of the Meetings of the AEC, Box 12, Meeting No. 1434, 2 December 1958, 875–77.

45. NACP, RG 326 Records of the Atomic Energy Commission, Entry A1 19, Minutes of the Meetings of the AEC, Box 12, Meeting No. 1434, 2 December 1958, 877.

46. Later on the same day, 2 December 1958, the AEC commissioners reconvened for another meeting and again discussed nuclear weapons tests. NACP, RG 326 Records of the Atomic Energy Commission, Entry A1 19, Minutes of the Meetings of the AEC, Box 12, Meeting No. 1435, 2 December 1958, 880–81.

47. NACP, RG 326 Records of the Atomic Energy Commission, Entry A1 19, Minutes of the Meetings of the AEC, Box 12, Meeting No. 1440, 9 December 1958, 909–11.

48. NACP, RG 326 Records of the Atomic Energy Commission, Entry A1 19, Minutes of the Meetings of the AEC, Box 2, Meeting no. 184, 8 July 1948, 165.

49. For example, see debate in these AEC meetings: NACP, RG 326 Records of the Atomic Energy Commission, Entry A1 19, Minutes of the Meetings of the AEC, Box 2, Notes on Conference, 2 September 1948, 4; Box 2, Meeting no. 221, 1 December 1948, 125–26.

50. Harry S. Truman Presidential Library, Independence, Missouri (hereafter HSTL), Papers of Truman, President's Secretary's Files, Box 174, Folder Atomic Bomb Reports, US Weather Bureau Report on Alert Number 122 of the Atomic Detection System, 29 September 1949, 1.

51. NACP, RG 326 Records of the Atomic Energy Commission, Entry A1 19, Minutes of the Meetings of the AEC, Box 1, Meeting no. 346, 21 December 1949, 491.

52. HSTL, Papers of Truman, President's Secretary's Files, Box 176, Folder Atomic Energy, Superbomb Data, Memo from President to Attorney General, 4 January 1951.

53. Joseph and Stewart Alsop, "How Red A-Blast Was Detected," *Washington Post*, 31 December 1950.

54. *Twenty-fourth Semiannual Report of the Atomic Energy Commission, July 1958*, 394.

55. NACP, RG 326 Records of the Atomic Energy Commission, Entry A1 19, Minutes of the Meetings of the AEC, Box 12, Meeting No. 1448, 22 December 1958, 993–95.

56. DDEL, John A. McCone Papers, Box 6, Folder Testing File #1 Eyes Only (3), Letter from H. S. Vance to Christian A. Herter, 23 December 1958.

57. DDEL, John A. McCone Papers, Box 6, Folder Testing File #1 Eyes Only (3), Letter from H. S. Vance to Christian A. Herter, 23 December 1958.

58. NACP, RG 326 Records of the Atomic Energy Commission, Entry A1 19, Minutes of the Meetings of the AEC, Box 12, Meeting No. 1459, 26 January 1959, 75–76.

59. NACP, RG 326 Records of the Atomic Energy Commission, Entry A1 19, Minutes of the Meetings of the AEC, Box 12, Meeting No. 1459, 26 January 1959, 76–78.

60. Barth, "Detecting the Cold War," 169–82.

61. DDEL, John A. McCone Papers, Box 6, Folder Testing File #1 Eyes Only (1), The Need for Fundamental Research in Seismology: Report of the Panel on Seismic Improvement, 31 March 1959, 1–2.

62. DDEL, John A. McCone Papers, Box 6, Folder Testing File #1 Eyes Only (1), Findings of the Panel on Seismic Improvement, 1–4.

63. DDEL, John A. McCone Papers, Box 6, Folder, Test File—March 1960, Letter Eisenhower to Khrushchev, 13 April 1959.

64. DDEL, John A. McCone Papers, Box 6, Folder, Test File—March 1960, Letter Khrushchev to Eisenhower, 23 April 1959.

65. NACP, RG 326 Records of the Atomic Energy Commission, Entry A1 19, Minutes of the Meetings of the AEC, Box 12, Meeting No. 1505, 8 May 1959, 316. NACP, RG 326 Records of the Atomic Energy Commission, Entry A1 19, Minutes of the Meetings of the AEC, Box 12, Meeting No. 1506, 12 May 1959, 319–20.

66. DDEL, Dwight D. Eisenhower Papers as President, Ann Whitman File, Press Conference Series, Box 8, Press Conference 17 June 1959, Official White House Transcript of President Eisenhower's Press and Radio Conference, #161, 3–4.

67. NACP, RG 326 Records of the Atomic Energy Commission, Entry A1 19, Minutes of the Meetings of the AEC, Box 13, Meeting No. 1526, 2 July 1959, 436–37.

68. On the first mention of the Interdepartmental Panel on Test Detection: NACP, RG 326 Records of the Atomic Energy Commission, Entry A1 19, Minutes of the Meetings of the AEC, Box 12, Meeting No. 1521, 19 June 1959. On all else: NACP, RG 326 Records of the Atomic Energy Commission, Entry A1 19, Minutes of the Meetings of the AEC, Box 13, Meeting No. 1526, 2 July 1959, 437–39.

69. One reason reaching an agreement remained unlikely stemmed from unresolved technical issues and the complicated decisions those necessitated. See, for example, White House Office, Office of the Staff Secretary, Records, Subject Series, Alphabetical Subseries, Box 16, Folder Dr. Kistiakowsky (1), Memo to President, 4 August 1959, 2–4.

70. DDEL, John A. McCone Papers, Box 6, Folder Testing File #2 Eyes Only (2), Memo Alfred D. Starbird to McCone, subj: AEC Position Relative to Testing, 26 October 1959.

71. NACP, RG 326 Records of the Atomic Energy Commission, Entry A1 19, Minutes of the Meetings of the AEC, Box 13, Meeting No. 1572, 11 December 1959, 663.

72. NACP, RG 326 Records of the Atomic Energy Commission, Entry A1 19, Minutes of the Meetings of the AEC, Box 13, Meeting No. 1573, 15 December 1959, 676.

73. NACP, RG 326 Records of the Atomic Energy Commission, Entry A1 19, Minutes of the Meetings of the AEC, Box 13, Meeting No. 1575, 19 December 1959, 692.

74. DDEL, John A. McCone Papers, Box 6, Folder, Test File—March 1960, Statement by the President, 11 February 1960.

75. DDEL, Dwight D. Eisenhower Papers as President, Ann Whitman File, Press

Conference Series, Box 10, Press Conference 17 February 1960, Briefing Paper, 1–2.

76. DDEL, Dwight D. Eisenhower Papers as President, Ann Whitman File, Press Conference Series, Box 10, Press Conference 17 February 1960, Official White House Transcript of President Eisenhower's Press and Radio Conference, #181, 7.

77. DDEL, Dwight D. Eisenhower Papers as President, Ann Whitman File, Press Conference Series, Box 10, Press Conference 27 April 1960, Briefing Paper, 1.

78. DDEL, Dwight D. Eisenhower Papers as President, Ann Whitman File, Press Conference Series, Box 10, Press Conference 27 April 1960, Official White House Transcript of President Eisenhower's Press and Radio Conference, #184, 9.

79. There is debate on this point within the historiography. For example, Per Fredrik Ilsaas Pharo, in his article "A Precondition for Peace," argued that the U-2 affair had little to do with a treaty never coalescing during the Eisenhower presidency, instead arguing that constant elements in the US government (and thus not external events) constituted the most important factors.

80. George B. Kistiakowsky, *A Scientist at the White House: The Private Diary of President Eisenhower's Special Assistant for Science and Technology* (Cambridge: Harvard University Press, 1976), 312–14, 321–22, 324.

81. NACP, RG 326 Records of the Atomic Energy Commission, Entry A1 19, Minutes of the Meetings of the AEC, Box 13, Meeting No. 1616, 3 May 1960, 320–23.

82. DDEL, Dwight D. Eisenhower Papers as President, Ann Whitman File, Press Conference Series, Box 10, Press Conference 11 May 1960, Briefing Paper.

83. DDEL, John A. McCone Papers, Box 4, Folder RES&D 1–2—TESTING—March 1960–July 1960 [Folder 1] (1), Memo from N. F. Twining To Secretary of Defense, 13 June 1960, subj: Draft Treaty on the Discontinuance of Nuclear Weapons Tests.

84. NACP, RG 326 Records of the Atomic Energy Commission, Entry A1 19, Minutes of the Meetings of the AEC, Box 13, Meeting No. 1634, 1 July 1960, 491–97.

85. NACP, 326 Records of the Atomic Energy Commission, Entry A1 19, Minutes of the Meetings of the AEC, Box 13, Meeting No. 1636, 6 July 1960, 502–5.

86. DDEL, John A. McCone Papers, Box 4, Folder RES&D 1–2—TESTING—March 1960–July 1960 [Folder 1] (1), Position Paper: Nuclear Test Suspensions High Altitude Detection Problem, 21 July 1960.

87. DDEL, Dwight D. Eisenhower Papers as President, Ann Whitman File, Press Conference Series, Box 10, Press Conference 17 August 1960, Briefing Papers.

88. Of course, the quotation continued to say: "regardless of whether or not the fears of fallout radiation are scientifically valid. Accordingly, the commission would propose that future tests, if any, be conducted underground, or in space beyond the earth's atmosphere, in such a way as not to cause fallout." *Annual Report to the Congress of the Atomic Energy Commission for 1960, January 1961*, xiii–xiv. A good brief summary of the negotiations is contained on pages 125–27.

89. Kistiakowsky, *A Scientist at the White House*, 375.

90. This point has been well debated in the historiography, and therefore this book

merely notes Eisenhower's beliefs, whether those were correct or not. For example, see Pharo, "A Precondition for Peace"; Smith-Norris, "The Eisenhower Administration and the Nuclear Test Ban Talks, 1958–1960"; Greene, *Eisenhower, Science Advice, and the Nuclear Test-Ban Debate, 1945–1963*.

CHAPTER FOUR

1. Joe Musial, *Learn How Dagwood Splits the Atom!* (King Features Syndicate, Inc., 1949), 31.

2. *Sixth Semiannual Report of the Atomic Energy Commission, July 1949*, 16.

3. Musial, *Learn How Dagwood Splits the Atom!*, introduction.

4. Joel B. Hagen, *An Entangled Bank: The Origins of Ecosystem Ecology* (New Brunswick: Rutgers University Press, 1992), 100–101.

5. In addition to this story at the AEC level, see nanoscientist Paige Johnson's research into atomic gardening. "Atomic Gardening: An Online History," http://www.atomicgardening.com. Accessed 26 June 2017.

6. *Time* 50, no. 5 (4 August 1947). In the Christian holy texts, the red horse rider of the Apocalypse represents war. Revelation 6:3–4 reads, "When the Lamb opened the second seal, I heard the second living creature say, 'Come and see!' Then another horse came out, a fiery red one. Its rider was given power to take peace from the earth and to make men slay each other. To him was given a large sword." *Bible* (New International Version).

7. As Edmund Russell has written, humans have yet to figure out "how to transform sunlight, carbon dioxide, and a few nutrients into grain—except by subcontracting the job to plants." Edmund Russell, "The Garden in the Machine: Toward an Evolutionary History of Technology," in *Industrializing Organisms: Introducing Evolutionary History*, ed. Susan R. Schrepfer and Philip Scranton (New York: Routledge, 2004), 8.

8. Donald Worster, "Transformations of the Earth: Toward an Agroecological Perspective in History." *Journal of American History* 76, no. 4 (March 1990), 1091.

9. Sterling Evans, "Agricultural Production and Environmental History," in *The Oxford Handbook of Food History*, Jeffrey M. Pilcher, ed. (Oxford University Press, 2012), 209–26. For an essay that uses lessons from environmental history to make agriculture more relatable (and visible) to modern-day peoples, see Sterling Evans, "The 'Age of Agricultural Ignorance': Trends and Concerns for Agriculture into the Twenty-First Century," *Agricultural History* 93, no. 1 (Winter 2019): 4–34.

10. Paul S. Sutter, "The World with Us: The State of American Environmental History," *Journal of American History* 100, no.1 (June 2013): 99, 105–9.

11. Peter A. Coclanis, "Field Notes: Agricultural History's New Plot," *Journal of Interdisciplinary History* 50, no. 2 (Autumn, 2019), 194–99.

12. These contexts continued to intertwine after 1960 as well shown by Jacob Darwin Hamblin, "Let there be light . . . and bread: The United Nations, the developing

world, and atomic energy's Green Revolution," *History and Technology* 25, no. 1 (March 2009).

13. One excellent example of both the technologies and mindsets involved in modernized agricultural production can be found in Melody Petersen, "As Beef Cattle become Behemoths, Who Are Animal Scientists Serving?" *Chronicle of Higher Education*, 15 April 2012. http://chronicle.com/article/As-Beef-Cattle-Become/131480/ accessed 9 June 2016.

14. Deborah Fitzgerald, *Every Farm a Factory: The Industrial Ideal in American Agriculture* (New Haven: Yale University Press, 2003).

15. David B. Danbom, *Born in the Country: A History of Rural America* (Baltimore: The Johns Hopkins University Press, 1995), 233–34.

16. One of the very best examinations of the development of the US chemical industry surely is Edmund Russell's study of pesticides and war gases that argues the two coevolved, especially in the rhetoric of their producers. Edmund Russell, *War and Nature: Fighting Humans and Insects with Chemicals from World War I to* Silent Spring (Cambridge: Cambridge University Press, 2001). In animal feed, see: chapter 5, "Modern Meat: Hormones in Livestock" in Nancy Langston, *Toxic Bodies: Hormone Disruptors and the Legacy of DES* (New Haven: Yale University Press, 2010), 61–82.

17. John H. Perkins, *Geopolitics and the Green Revolution: Wheat, Genes, and the Cold War* (New York: Oxford University Press, 1997).

18. J. L. Anderson, *Industrializing the Corn Belt: Agriculture, Technology, and Environment, 1945–1972* (DeKalb: Northern Illinois University Press, 2009); David Vail, *Chemical Lands: Pesticides, Aerial Spraying, and Health in North America's Grasslands since 1945* (Tuscaloosa: University of Alabama Press, 2018).

19. On family farming becoming a business see John Fraser Hart, *The Land That Feeds Us* (New York: W. W. Norton & Company, 1991), 356. On the development of cross-country trucking networks, see Shane Hamilton, *Trucking Country: The Road to America's Wal-Mart Economy* (Princeton: Princeton University Press, 2008), 5, 58.

20. US Bureau of Labor Statistics, "100 Years of US Consumer Spending: Data for the Nation, New York City, and Boston" (Report 991, May 2006), 3

21. Mike Davis, *Late Victorian Holocausts: El Niño Famines and the Making of the Third World* (New York: Verso, 2002).

22. Nick Cullather, *The Hungry World: America's Cold War Battle against Poverty in Asia* (Cambridge: Harvard University Press, 2010), ix, 7.

23. Agricultural surpluses were not a huge national problem during Truman's administration, but after US involvement in the Marshall Plan and the Korean War both ended these surpluses became significant policy problems during Eisenhower's presidency and later. President Lyndon Johnson especially used PL-480 to help spread his Great Society program all across the world with the explicit mission of helping feed global poor. As Johnson said in a speech at the University of Kentucky in February 1965,

"We care that men are hungry—not only in Appalachia but in Asia and Africa and in other spots of the world." Kristin L. Ahlberg, *Transplanting the Great Society: Lyndon Johnson and Food for Peace* (Columbia: University of Missouri Press, 2008), 4, 13–20.

24. Angus Wright, *The Death of Ramón González: The Modern Agricultural Dilemma* (Austin: University of Texas Press, 1990).

25. For a broader history of radioisotope tracers, see Angela N. H. Creager, *Life Atomic: A History of Radioisotopes in Science and Medicine* (Chicago: University of Chicago Press, 2013).

26. Harry S. Truman Presidential Library, Independence, Missouri (hereafter HSTL), Papers of Truman, President's Secretary's Files, Box 174, Folder Atomic Bomb, Press Releases [2 of 3], Press Release "First Peacetime Application of Atomic Research Becomes Immediately Possible under Army Program," 14 June 1946.

27. Italicized text in this speech appears as underlined text in the original document. HSTL, Papers of Clark Clifford, Subject File, 1945–54, Box 1, Folder Atomic Energy—Lilienthal, Atomic Energy is Your Business, 22 September 1947, 1–3.

28. Lilienthal made this comparison even though a sun is powered by nuclear fusion and the chain reactions humans could create were nuclear fission. A similar comparison was made here: HSTL, Papers of Truman, President's Secretary's Files, Box 194, Folder ABomb War Department, Atomic Energy Source of Inexhaustible Power

29. HSTL, Papers of Clark Clifford, Subject File, 1945–54, Box 1, Folder Atomic Energy—Lilienthal, Atomic Energy is *Your* Business, 22 September 1947, 4–5.

30. Particularly, Lilienthal claimed that engineering knowledge was not needed to make decisions about the atom, but instead, "what is needed is sense about human relations, about standards of fairness, about principles of self-government and of self-education" so that "petty politics" would not get in the way of progress. HSTL, Papers of Clark Clifford, Subject File, 1945–54, Box 1, Folder Atomic Energy—Lilienthal, Atomic Energy is Your Business, 22 September 1947, 5–9.

31. Jacob Darwin Hamblin, *Arming Mother Nature: The Birth of Catastrophic Environmentalism* (New York: Oxford University Press, 2013), 33, 40.

32. Historian Donald Worster wrote that for agriculture around the turn of the twentieth century, the "most important new wrinkle was the machine." Donald Worster, *Dust Bowl: The Southern Plains in the 1930s* (New York: Oxford University Press, 1979), 87.

33. HSTL, Papers of Clark Clifford, Subject File, 1945–54, Box 1, Folder Atomic Energy—Lilienthal, Atomic Energy and the American Farmer, 16 December 1947, 1.

34. Italicized text in this passage appears as underlined text in the original document. HSTL, Papers of Clark Clifford, Subject File, 1945–54, Box 1, Folder Atomic Energy—Lilienthal, Atomic Energy and the American Farmer, 16 December 1947, 1–4.

35. The religious undertones were not uncommon at the time, either. In January 1951, one concerned citizen wrote to "Mrs. 'First Lady'" to protest nuclear tests in the desert as being against divine plan. "This area which is chosen for the sake of

proving the effectiveness of atomic destruction, was created by God to demonstrate to man the miraculous power of construction. Proof of this fact is God's Law, which says: 'The desert shall bud and blossom as the rose.'" The civilian also wrote, "May I add: that you secure about 2 bushels of desert top sand from Nevada and I will gladly show you how to have most beautiful flowers," and claimed that, with more nuclear tests, "our farm problems are most serious now." HSTL, Papers of Harry S. Truman, Official File, Box 1524, Folder 692-Misc. (1950–53), Letter Hamilton L. Roe to Mrs. H. S. Truman, 12 January 1951.

36. HSTL, Papers of Clark Clifford, Subject File, 1945–54, Box 1, Folder Atomic Energy—Lilienthal, Atomic Energy and the American Farmer, 16 December 1947, 2, 5–6.

37. Italicized text in this passage appears as underlined text in the original document. HSTL, Papers of Clark Clifford, Subject File, 1945–54, Box 1, Folder Atomic Energy—Lilienthal, Atomic Energy and the American Farmer, 16 December 1947, 6, 9, 11, 12.

38. HSTL, Papers of Clark Clifford, Subject File, 1945–54, Box 1, Folder Atomic Energy—Lilienthal, Atomic Energy and the American Farmer, 16 December 1947, 13–16.

39. Cullather, *The Hungry World*, 4.

40. Jenny Barker-Devine, "'Mightier Than Missiles': The Rhetoric of Civil Defense for Rural American Families, 1950–1970." *Agricultural History* 80, no. 4 (Autumn, 2006), 415.

41. The National Archives at College Park, Maryland (hereafter NACP), RG 326 Records of the Atomic Energy Commission, Entry A1 19, Minutes of the Meetings of the AEC, Box 2, Meeting no. 218, 23 November 1948, 115.

42. *Fifth Semiannual Report of the Atomic Energy Commission, January 1949*, 90–91.

43. *Sixth Semiannual Report of the Atomic Energy Commission, July 1949*, 18, 21, 101, 104.

44. *Seventh Semiannual Report of the Atomic Energy Commission, January 1950*, 21–23.

45. In general, the AEC tried to downplay any potential dangers of nuclear energy, especially radiation. An October 1950 pamphlet on "Survival Under Atomic Attack" contained a section on "Kill the Myths." Those "wrong" myths were (1) "Atomic *weapons* will not destroy the earth"; (2) "Doubling bomb power does not double destruction"; (3) "Radioactivity is not the bomb's greatest threat"; (4) "Radiation sickness is not always fatal." HSTL, Papers of Harry S. Truman, Official File, Box 1527, Folder OF 692-A, Atomic Bomb, Survival Under Atomic Attack, 29 October 1950.

46. *Eighth Semiannual Report of the Atomic Energy Commission, July 1950*, 172.

47. *Ninth Semiannual Report of the Atomic Energy Commission, January 1951*, 24.

48. *Ninth Semiannual Report of the Atomic Energy Commission, January 1951*, 24–25, 28.

49. *Tenth Semiannual Report of the Atomic Energy Commission, July 1951*, 42–44.

50. Inflation statistics cited from the inflation calculator on the US Bureau of Labor Statistics' Consumer Price Index., http://data.bls.gov/cgi-bin/cpicalc.pl.htm. Accessed 18 July 2019.

51. *Eleventh Semiannual Report of the Atomic Energy Commission, January 1952,* 69, 71, 75.

52. *Eleventh Semiannual Report of the Atomic Energy Commission, January 1952,* 75, 81.

53. John Hersey, *Hiroshima* (New York: Bantam Books, Inc., 1946, 1981), 69–70.

54. *Eleventh Semiannual Report of the Atomic Energy Commission, January 1952,* 82–83.

55. Not only did *Hiroshima* wake those in the United States to atomic bombs' potentially horrific effects, it also presented the natural world as a counter to that destruction: "All day, people poured into Asano Park. This private estate was far enough away from the explosion so that its bamboos, pines, laurel, and maples were still alive, and the green place invited refugees—partly because they believed that if the Americans came back, they would bomb only buildings; partly because the foliage seemed a center of coolness and life, and the estate's exquisitely precise rock gardens, with their quiet pools and arching bridges, were very Japanese, normal, secure; and also partly (according to some who were there) because of an irresistible atavistic urge to hide under the leaves." Hersey, *Hiroshima,* 46.

56. *Eleventh Semiannual Report of the Atomic Energy Commission, January 1952,* 101.

57. In biological processes, Sr^{90} mimics calcium and therefore plants readily draw in the fallout product. Once Sr^{90} enters plant tissues it then bioaccumulates as it meanders up the food chain and eventually gets passed onto humans, especially in cow's milk (the largest source of calcium in the human diet).

58. *Eleventh Semiannual Report of the Atomic Energy Commission, January 1952,* 86–89, 92.

59. *Eleventh Semiannual Report of the Atomic Energy Commission, January 1952,* 93, 95, 99, 101, 123–26.

60. Dwight D. Eisenhower Presidential Library, Abilene, Kansas (hereafter DDEL), Dwight D. Eisenhower Papers as President, Ann Whitman File, Campaign Series, Box 6, Folder Atomic Energy, "General Principles Regarding Atomic Energy Development," 1–3.

61. Edward L. Schapsmeier and Frederick H. Schapsmeier, "Eisenhower and Agricultural Reform: Ike's Farm Policy Legacy Appraised," *American Journal of Economics and Sociology* 51, no. 2 (Apr. 1992), 153.

62. *Thirteenth Semiannual Report of the Atomic Energy Commission, January 1953,* 120.

63. *Fifteenth Semiannual Report of the Atomic Energy Commission, January 1954,* 46.

64. DDEL, White House Central Files, Official File, Box 449, Folder 1, Atomic Energy and the Improvement of Agriculture, 12 January 1954, 1–2.

65. DDEL, White House Central Files, Official File, Box 449, Folder 1, Atomic Energy and the Improvement of Agriculture, 12 January 1954, 2–4.

66. DDEL, White House Central Files, Official File, Box 449, Folder 1, Atomic Energy and the Improvement of Agriculture, 12 January 1954, 5–6.

67. DDEL, White House Central Files, Official File, Box 449, Folder 1, Atomic Energy and the Improvement of Agriculture, 12 January 1954, 6–7.

68. DDEL, White House Central Files, Official File, Box 449, Folder 1, Atomic Energy and the Improvement of Agriculture, 12 January 1954, 7.

69. *Sixteenth Semiannual Report of the Atomic Energy Commission, July 1954*, 56–58.

70. NACP, 326 Records of the Atomic Energy Commission, Entry 73, Division of Biology and Medicine, Records Relating to Fallout Studies, 1953–6, Box 8, Folder USDA (Beltsville)—Soil Program (Sunshine), 1949–1957.

71. DDEL, White House Central Files, Official File, Box 212, Folder 6, Letter Michael V. DiSalle to James Hagerty, 8 December 1954.

72. *Seventeenth Semiannual Report of the Atomic Energy Commission, January 1955*, 48.

73. *Eighteenth Semiannual Report of the Atomic Energy Commission, July 1955*, 85–86.

74. *Nineteenth Semiannual Report of the Atomic Energy Commission, January 1956*, 78–81.

75. DDEL, White House Central Files, Official File, Box 449, Folder 5, Letter DDE to Carl Hinshaw, 3 September 1954.

76. DDEL, Dwight D. Eisenhower Papers as President, Ann Whitman File, Administrative Series, Box 9, Folder Operation "Candor" (1), Report of the Panel on the Impact of the Peaceful Uses of Atomic Energy to the Joint Committee on Atomic Energy, Volume 1, January 1956, 61.

77. As Nick Cullather explained, the Green Revolution led to a new type of international politics: "How and on what terms Asia's population would be integrated into the world economy, whether fragile postcolonial states could extend mechanisms of taxation and authority over vast ungoverned hinterlands, and whether poverty on this scale even could be ameliorated were all questions that lay outside of the customary conventions of international relations." Cullather, *The Hungry World*, 3–4.

78. DDEL, Dwight D. Eisenhower Papers as President, Ann Whitman File, Administrative Series, Box 9, Folder Operation "Candor" (1), Report of the Panel on the Impact of the Peaceful Uses of Atomic Energy to the Joint Committee on Atomic Energy, Volume 1, January 1956, 61.

79. On plant breeding from a scientific perspective using the concept of large technical systems, see: Helen Anne Curry, "Atoms in Agriculture," *Historical Studies in the Natural Sciences* 46, no. 2 (April 2016): 119–53.

80. As quoted in Jacob Darwin Hamblin, *The Wretched Atom: America's Global Gamble with Peaceful Nuclear Technology* (Oxford University Press, 2021), 3.

81. Some useful variations had already been created using induced radiation

mutations. Those listed were (1) barley, "dense heads, stiff straw, tall straw, higher yield of grain and straw"; (2) oats, "earliness, higher yield, stem-rust resistance, short stems"; (3) wheat, "stem-rust resistance, higher yield"; (4) corn, "shorter or taller stalks, earlier or later ripening, resistance to lodging"; and (5) peanuts, "leaf-spot resistance, higher yield." DDEL, Dwight D. Eisenhower Papers as President, Ann Whitman File, Administrative Series, Box 9, Folder Operation "Candor" (1), Report of the Panel on the Impact of the Peaceful Uses of Atomic Energy to the Joint Committee on Atomic Energy, Volume 1, January 1956, 64.

82. DDEL, Dwight D. Eisenhower Papers as President, Ann Whitman File, Administrative Series, Box 9, Folder Operation "Candor" (1), Report of the Panel on the Impact of the Peaceful Uses of Atomic Energy to the Joint Committee on Atomic Energy, Volume 1, January 1956, 65.

83. DDEL, Dwight D. Eisenhower Papers as President, Ann Whitman File, Administrative Series, Box 9, Folder Operation "Candor" (1), Report of the Panel on the Impact of the Peaceful Uses of Atomic Energy to the Joint Committee on Atomic Energy, Volume 1, January 1956, 65–67. For the logical endpoint of these ideas that came to see supermarkets as "weapons" in the Cold War Farms Race, as historian Shane Hamilton has called it, see Hamilton, *Supermarket USA: Food and Power in the Cold War Farms Race* (New Haven: Yale University Press, 2018). Previously, Hamilton has also interpreted these surpluses, and the trucks used to transport the foodstuffs around the country, as aggressively undermining New Deal liberalism with free market solutions. Hamilton argued, "Trucks . . . were inherently political technologies, used by agribusinesses to craft 'free market' solutions to the farm problem while ironically allowing regressive New Deal farm policies to outlive the labor, consumer, and regulatory programs of the New Deal." Hamilton, *Trucking Country*, 7.

84. DDEL, Dwight D. Eisenhower Papers as President, Ann Whitman File, Administrative Series, Box 9, Folder Operation "Candor" (1), Report of the Panel on the Impact of the Peaceful Uses of Atomic Energy to the Joint Committee on Atomic Energy, Volume 1, January 1956, 67–68.

85. "Atoms Vital to Agriculture," *Science News Letter* 69, no. 3 (21 January 1956): 35.

86. *Twentieth Semiannual Report of the Atomic Energy Commission, July 1956*, xi.

87. *Twenty-First Semiannual Report of the Atomic Energy Commission, January 1957*, 80.

88. DDEL, White House Central Files, Official File, Box 453, Folder OF 108-F Atoms for Peace (5), Material prepared for hearings by the Congressional Joint Committee on Atomic Energy, February 1957, 1–2, 13–14, 20.

89. *Twenty-second Semiannual Report of the Atomic Energy Commission, July 1957*, 29, 32–33.

90. *Twenty-second Semiannual Report of the Atomic Energy Commission, July 1957*, 116–17.

91. The AEC claimed that in saving about $500 million per year, radioisotopes

produced a better than 7 percent yearly return on the $7 billion spent on them in tax-payer money. *Twenty-third Semiannual Report of the Atomic Energy Commission, January 1958*, 4.

92. The soil moisture gauges worked, in layman's terms, by zipping radiation into the soil and then measuring what got reflected back into the gauge. *Twenty-third Semiannual Report of the Atomic Energy Commission, January 1958*, 26–30.

93. Italics are in the original. *Twenty-third Semiannual Report of the Atomic Energy Commission, January 1958*, 36.

94. These two chemicals, combined in a 50:50 mixture with an organic solvent like diesel fuel, formed the active ingredients in the infamous Agent Orange used so pro-fusely during the Vietnam War. *Twenty-third Semiannual Report of the Atomic Energy Commission, January 1958*, 57–63.

95. The January 1958 AEC report to Congress listed some of the particularly bene-ficial mutations derived from mutations to be disease-resistant strains in wheat, oats, and flax, and high-yield dwarf varieties that suffer from less wind damage. *Twenty-third Semiannual Report of the Atomic Energy Commission, January 1958*, 64–67.

96. NACP, RG 326 Records of the Atomic Energy Commission, Entry A1 19, Min-utes of the Meetings of the AEC, Box 13, Meeting No. 1595, 25 February 1960, 152–53.

97. Moreover, their experiments "continued to demonstrate whether foods ster-ilized by this process would be nontoxic, acceptable to consumers, adaptable to ef-ficient handling techniques, and economically competitive with foods preserved by other processes." *Twentieth Semiannual Report of the Atomic Energy Commission, July 1956*, 52. Kaiser Engineers from Oakland, California eventually won the bidding pro-cess "to design and construct a Food Irradiation Reactor (FIR) for the Army's Ioniz-ing Radiation Center." *Twenty-First Semiannual Report of the Atomic Energy Commission, January 1957*, 52.

98. McCone's sentiments were paraphrased in the notes, and therefore it is unlikely that the wording is a direct quotation of his. NACP, RG 326 Records of the Atomic En-ergy Commission, Entry A1 19, Minutes of the Meetings of the AEC, Box 13, Meeting No. 1595, 25 February 1960, 153–54.

99. NACP, RG 326 Records of the Atomic Energy Commission, Entry A1 19, Min-utes of the Meetings of the AEC, Box 13, Meeting No. 1603, 1 April 1960, 221–23.

100. From Eisenhower's address at Bradley University, Peoria, Illinois on 25 Sep-tember 1956, http://www.eisenhower.archives.gov. Accessed on 21 August 2014.

101. DDEL, White House Central Files, Official File, Cross Reference Sheets, Box 88, Folder OF 108 Atomic Energy 1960, Cross Reference Sheet, Memo for the Re-cord, 24 June 1960.

102. DDEL, Dwight D. Eisenhower Papers as President, Ann Whitman File, Ad-ministrative Series, Box 5, Folder Atomic Energy Commission 1960–61, Letter Mc-Cone to Eisenhower, 3 January 1961.

103. Hamblin, "Let There Be Light ... and Bread," *History and Technology* 40, no. 42.

104. Keith Meyers, "In the Shadow of the Mushroom Cloud: Nuclear Testing, Radioactive Fallout, and Damage to U.S. Agriculture, 1945–1970," *The Journal of Economic History* 79, no. 1 (March 2019): 268–70.

CHAPTER FIVE

1. Conant was former Harvard chemistry professor and at the time of this speech president of the university. Harry S. Truman Presidential Library, Independence, Missouri (hereafter HSTL), Papers of Clark Clifford, Box 41, Folder Conant, J.B. ["The Atomic Age," a lecture at the National War College, October 2, 1947.]

2. Jacob Darwin Hamblin, "Gods and Devils in the Details: Marine Pollution, Radioactive Waste, and an Environmental Regime circa 1972," *Diplomatic History* 32, no. 4 (September 2008): 540. On larger ideas of abundance and scarcity, and on how human views of the American environment have shifted among those ideas, see Donald Worster, *Shrinking the Earth: The Rise and Decline of American Abundance* (New York: Oxford University Press, 2016). And on "the practices that coproduced global knowledge and international institutions" within the context of the global origins of sustainable-development policy, see Perrin Selcer, *The Postwar Origins of the Global Environment: How the United Nations Built Spaceship Earth* (Columbia University Press, 2018), 8.

3. In fact, a section in the July 1950 AEC Report to Congress described "Dilution as a Safeguard" when discussing airborne nuclear waste. *Eighth Semiannual Report of the Atomic Energy Commission, July 1950,* 80.

4. Jacob Darwin Hamblin, *Poison in the Well: Radioactive Waste in the Oceans at the Dawn of the Nuclear Age* (New Brunswick: Rutgers University Press, 2008), 2.

5. This sentiment is similar to when Dolly Jørgensen argued, "It is impossible to separate water use, industrial pollution and garbage disposal from the technologies that make them possible." Dolly Jørgensen, "Not by Human Hands: Five Technological Tenets for Environmental History in the Anthropocene," *Environment and History* 20, no. 4 (November 2014), 480.

6. Allison Macfarlane, "Underlying Yucca Mountain: The Interplay of Geology and Policy in Nuclear Waste Disposal," *Social Studies of Science* 33, no. 5, Earth Sciences in the Cold War (Oct. 2003): 783–807.

7. Joel Tarr's study of Pittsburgh from 1880–2000 reaches many similar conclusions, though it is perhaps more sanguine about humans' ability to sustainably build civilizations in terms of industrial pollution due to its end date four decades after this book. Joel A. Tarr, "The Metabolism of the Industrial City: The Case of Pittsburgh," in *City, Country, Empire: Landscapes in Environmental History,* ed. Jeffry M. Diefendorf and Kurk Dorsey (Pittsburgh: University of Pittsburgh Press, 2005), 15–37.

8. See, for example, chapter 3, "Reactors," in Angela N. H. Creager, *Life Atomic: A History of Radioisotopes in Science and Medicine* (The University of Chicago Press, 2013).

9. Darwin's source is a 1951 *New York Times* article titled "Atomic Death Belt Urged

for Korea." Jacob Darwin Hamblin, *Arming Mother Nature: The Birth of Catastrophic Environmentalism* (New York: Oxford University Press, 2013), 48–49.

10. *Fifth Semiannual Report of the Atomic Energy Commission, January 1949*, 4.

11. *Sixth Semiannual Report of the Atomic Energy Commission, July 1949*, 62.

12. In December 1949, the AEC commissioners discussed commissioning such a study. The National Archives at College Park, Maryland (hereafter NACP), RG 326 Records of the Atomic Energy Commission, Entry A1 19, Minutes of the Meetings of the AEC, Box 1, Meeting No. 343, 8 December 1949, 379.

13. The AEC report six months later in January 1950 contained a mere paragraph on "Handling Radioactive Wastes" that briefly mentioned a somewhat inconsequential study at Mound Laboratory at Miamisburg, Ohio. *Seventh Semiannual Report of the Atomic Energy Commission, January 1950*, 124.

14. *Eighth Semiannual Report of the Atomic Energy Commission, July 1950*, 98.

15. *Eighth Semiannual Report of the Atomic Energy Commission, July 1950*, 98–99.

16. The AEC seemed confident that testing showed no appreciable environmental harm from radioactivity. On Oak Ridge, the commission claimed, "The U.S. Weather Bureau conducts thorough meteorological surveys under a cooperative agreement with the AEC. The work involves detailed measurement of wind flow and temperature in the broken ridge country surrounding the plants, including the behavior of the upper atmosphere." About Argonne, the commission asserted, "Here, as at other atomic energy centers, tests are conducted to make sure that plant, fish, and animal life downstream from the Laboratory are not reconcentrating extremely dilute radioactive wastes to any dangerous extent." *Eighth Semiannual Report of the Atomic Energy Commission, July 1950*, 99–105.

17. The AEC was relatively secure in its assertions that it had everything under control, such as when the report stated, "From evidence now at hand there is no reason to believe that the operation of the Hanford piles will have any harmful effect on the natural balance of plant and animal life in the [Columbia] river." *Eighth Semiannual Report of the Atomic Energy Commission, July 1950*, 105–18.

18. *Tenth Semiannual Report of the Atomic Energy Commission, July 1951*, 42.

19. Or, to put it within the framework proposed by historian of science Thomas Kuhn, these policymakers could not break out of their existing paradigm. Thomas S. Kuhn, *The Structure of Scientific Revolutions* (Chicago: University of Chicago Press, 1962, 2012).

20. *Twelfth Semiannual Report of the Atomic Energy Commission, July 1952*, 19–20. Various nuclear refinement processes produced radioactive "ash," which also could cause concern. Dwight D. Eisenhower Presidential Library, Abilene, Kansas (hereafter DDEL), Dwight D. Eisenhower Papers as President, Ann Whitman File, Campaign Series, Box 6, Folder Atomic Energy, Memo to Bernard M. Baruch from John M. Hancock, 29 September 1952, 4.

21. *Seventeenth Semiannual Report of the Atomic Energy Commission, January 1955*, 53.

22. *Fifteenth Semiannual Report of the Atomic Energy Commission, January 1954*, 26.

23. Other "methods for the disposal of contaminated scrap include storage, burial, sea disposal and incineration." *Seventeenth Semiannual Report of the Atomic Energy Commission, January 1955*, 30–31, 53.

24. *Nineteenth Semiannual Report of the Atomic Energy Commission, January 1956*, 51–52.

25. *Twentieth Semiannual Report of the Atomic Energy Commission, July 1956*, 60–62.

26. *Twenty-First Semiannual Report of the Atomic Energy Commission, January 1957*, 151–52.

27. *Twenty-First Semiannual Report of the Atomic Energy Commission, January 1957*, 156–58.

28. *Twenty-second Semiannual Report of the Atomic Energy Commission, July 1957*, 69–70.

29. *Twenty-third Semiannual Report of the Atomic Energy Commission, January 1958*, 133–51.

30. *Twenty-fourth Semiannual Report of the Atomic Energy Commission, July 1958*, 101–102.

31. Italics are in original. *Twenty-fourth Semiannual Report of the Atomic Energy Commission, July 1958*, 103.

32. The January 1959 AEC report to Congress stated, "An application for a license to dispose of radioactive waste in the ocean must include a detailed description of the quantities and kinds of material and the proposed manner and conditions of disposal. The applicant must give detailed information on container and packing specifications, processing facilities, transportation, instrumentation for measuring radiation levels and contamination, radiation safety procedures to be followed in collecting, storing, packaging, and transporting the material; site of disposal, including depth of water at the proposed site, and the records of disposal that will be maintained." *Twenty-fifth Semiannual Report of the Atomic Energy Commission, January 1959*, 13–14.

33. Edward Gamarekian, "HEW Acts to End A-Contamination of Rivers," *Washington Post and Times Herald*, 15 July 1959, A7.

34. NACP, RG 326 Records of the Atomic Energy Commission, Entry A1 19, Minutes of the Meetings of the AEC, Box 13, Meeting No. 1528, 15 July 1959, 446–47.

35. DDEL, White House Central Files, General File, Box 1214, Folder 155, 1959–60, Telegram from Robert G. MacLaughlin of Westkingston, RI to the President, 17 July 1959.

36. DDEL, White House Central Files, Official File, Box 450, Folder 4, Letter Christopher Del Sesto to President, 20 July 1959.

37. DDEL, White House Central Files, Official File, Box 450, Folder 4, Letter William B. Persons to Christopher Del Sesto, 3 August 1959.

38. DDEL, White House Central Files, General File, Box 1214, Folder 155, 1959–60, Telegram from Robert G. MacLaughlin of Westkingston, RI to the President, 17 July 1959.

39. DDEL, Dwight D. Eisenhower Papers as President, Ann Whitman File, Diary Series, Box 42, Folder Toner Notes July 1959, Special Legislative Note, 25 July 1959.

40. National Academy of Sciences—National Research Council, Publication 655, 1959, "Radioactive Waste Disposal into Atlantic and Gulf Coast Waters," i–vii.

41. It clarified this larger recommendation by saying before any final site was selected, five conditions had to be met: (1) a survey of the area must be conducted "to determine details of local circulation and an inventory of the biota, especially of bottom-living organisms"; (2) only so much could be dumped (radioactivity wise); (3) sites needed to be at least seventy-five miles apart and no three hundred mile stretch of coastline should have more than three sites; (4) containers should be designed so that if a container breaks no part of it should float to the top; (5) area should be monitored periodically. National Academy of Sciences—National Research Council, Publication 655, 1959, "Radioactive Waste Disposal into Atlantic and Gulf Coast Waters," 1–3.

42. As a comparison, the Ukrainian Chernobyl disaster released somewhere between hundreds of millions and billions of curies, depending on the estimate. From Nukewatch's Chernobyl factsheet, "Chernobyl: How much radiation was released?" http://nukewatchinfo.org. Accessed on 29 June 2017.

43. In one place, the report claimed: "Although impossible to evaluate quantitatively, it seems certain that sorption processes will play a major role in controlling the dispersal of radioactive wastes once they are free of the canister." In the end, the committee did not even include sorption factors into their calculations because they could not calculate it, but believed that this meant recommendations should "include a safety factor of at least 10, and possibly more." Clearly, in some instances they were just guessing. National Academy of Sciences—National Research Council, Publication 655, 1959, "Radioactive Waste Disposal into Atlantic and Gulf Coast Waters," 4–35.

44. The AEC thought that states could acceptably dispose of wastes in a variety of ways: burial, limited disposal into sewers, river dumping in low concentrations, incineration, ocean disposal, or even transfer to another licensed disposal agency. Or, states could reuse the products, and it was stated, "A limited number of licenses have been issued for studies involving the controlled release of radioisotopes into the environs. Examples of such field uses include fluid flow studies in oil wells and in streams. The quantities and dilutions involved in most field studies usually provide for radiation concentrations which are sufficient for technical measurements but which are a very small fraction of permissible levels." DDEL, White House Central Files, Official File, Box 454, Folder OF 108-F Atoms For Peace (12), Proposed Criteria for Guidance of States and the AEC in the Discontinuance of AEC Regulatory Authority Over Byproduct, Source and Special Nuclear Materials in Less Than a Critical Mass and the Assumption Thereof by States Through Agreement. The act

itself said its first purpose was "to recognize the interests of the States in the peaceful uses of atomic energy, and to clarify the respective responsibilities under this Act of the States and the Commission with respect to the regulation of byproduct, source, and special nuclear materials." DDEL, White House Central Files, Official File, Box 454, Folder OF 108-F Atoms For Peace (12), Public Law 86–373, 86th Congress, S. 2568, 23 September 1959.

45. DDEL, White House Office, Office of the Special Assistant, OCB Series, Subject Subseries, Box 5, Folder Nuclear Energy Matters (8) [Sept 1959-Mar 1960], Letter R. R. Rubottom, Jr. to John A. McCone, 18 November 1959. Rubottom was Assistant Secretary of State.

46. DDEL, White House Office, Office of the Special Assistant, OCB Series, Subject Subseries, Box 5, Folder Nuclear Energy Matters (8) [Sept 1959–Mar 1960], Memorandum for the Executive Officer, Subj: Industrial Radioactive Waste Disposal in the Gulf of Mexico 22 December 1959. By Richard Hirsch.

47. However, the memo continued, "Since the method of disposal is considered safe, the licensee may seek a judicial review. The [Operations Coordinating Board] has been alerted to the situation, and will attempt to deal with problems as they arise." DDEL, White House Office, Office of the Special Assistant, OCB Series, Administrative Subseries, Box 2, Folder Chronological—Karl G. Harr January 1960–January 1961 (1), Memo from J. I. Coffey to Toner, 5 January 1960.

48. NACP, RG 326 Records of the Atomic Energy Commission, Entry A1 19, Minutes of the Meetings of the AEC, Box 13, Meeting No. 1573, 15 December 1959, 670.

49. NACP, 326 Records of the Atomic Energy Commission, Entry A1 19, Minutes of the Meetings of the AEC, Box 13, Summary Notes of Meeting with Representatives of the State of New Jersey, 11 Feb 1960.

50. The study did conclude that, "However, highly radioactive wastes from plants located near the coast and requiring concrete shielding probably could be disposed of more cheaply at sea." NACP, RG 326 Records of the Atomic Energy Commission, Entry A1 19, Minutes of the Meetings of the AEC, Box 13, Meeting No. 1617, 6 May 1960, 325–27.

51. NACP, 326 Records of the Atomic Energy Commission, Entry A1 19, Minutes of the Meetings of the AEC, Box 14, Meeting No. 1675, 23 November 1960, 878–79.

52. Macfarlane, "Underlying Yucca Mountain."

53. On top of disposal at these facilities, the AEC also licensed nine total companies, four for the first time that year, all for "disposal of low-level waste commercially." *Annual Report to Congress of the Atomic Energy Commission for 1959, January 1960*, 124, 304–6.

54. *Annual Report to Congress of the Atomic Energy Commission for 1959, January 1960*, 306–8.

55. One of the more interesting maps of Hanford is located on the back of a letter sent to President Eisenhower. It not only shows Hanford situated between the Yakima

and Columbia Rivers but also includes lots of agricultural drawings, such as tractor, barn, wheat, cattle, fruit, etc. In addition, the map also portrays the Phillips Chemical Anhydrous Ammonia plant, the Boise-Cascade Co. Container Plant Pulp & Paper Mill, a Junior College, dams, and oil pipelines. In this way, the map depicts a melding of natural and humanmade parts of the local environment very reminiscent of Richard White's *Organic Machine*, also about the Columbia River. Obviously this map came decades before White's book. DDEL, White House Central Files, General File, Box 1214, Folder 155, 1958, Map of Hanford. Richard White, *The Organic Machine: The Remaking of the Columbia River* (New York: Hill and Wang, 1995).

56. HSTL, Papers of Truman, President's Secretary's Files, Box 174, Folder Atomic Bomb, Press Releases [1 of 3], War Department Press Release on Hanford Engineer Works, 1–2.

57. *Annual Report to Congress of the Atomic Energy Commission for 1959, January 1960*, 308–12.

58. DDEL, White House Central Files, General File, Box 1214, Folder 155, 1958, Letter from Donald A. Pugnetti to James Hagerty, 16 August 1958.

59. Creager, *Life Atomic*, 365–77.

60. *Annual Report to Congress of the Atomic Energy Commission for 1959, January 1960*, 313–17.

61. For a brief history of Hanford as place, especially in the years after this study, one good resource is: Melvin R. Adams, *Atomic Geography: A Personal History of the Hanford Nuclear Reservation* (Pullman: Washington State University Press, 2016).

62. An Imhoff tank is, in rough terms, a two-chamber septic tank so that the raw effluent is held in a separate chamber from the septic sludge. *Annual Report to Congress of the Atomic Energy Commission for 1959, January 1960*, 317–20.

63. *Annual Report to Congress of the Atomic Energy Commission for 1959, January 1960*, 321–29.

64. *Annual Report to Congress of the Atomic Energy Commission for 1959, January 1960*, 335–39. On the other types of research, see 339–67.

65. *Annual Report to Congress of the Atomic Energy Commission for 1960, January 1961*, 86.

66. Russell J. Dalton, Paula Garb, Nicholas P. Lovrich, John C. Pierce, John M. Whiteley, *Critical Masses: Citizens, Nuclear Weapons Production, and Environmental Destruction in the United States and Russia* (Cambridge: Massachusetts Institute of Technology Press, 1999), 6–7.

67. Michele A. Stenehjem, "Pathways of Radioactive Contamination: Beginning the History, Public Enquiry, and Policy Study of the Hanford Nuclear Reservation," *Environmental Review* 13, no. 3/4 (Autumn–Winter, 1989).

68. Andrew Jenks, "Model City USA: The Environmental Cost of Victory in World War II and the Cold War" *Environmental History* 12, no. 3 (Jul. 2007): 552.

69. Victor Gilinsky, "Yucca Mountain redux," *Bulletin of the Atomic Scientists*, 5

November 2014, http://thebulletin.org/yucca-mountain-redux7800. Accessed 29 December 2014.

70. Sarah Zhang, "America's Nuclear Waste Plan Is a Giant Mess: An Explosion Caused by Cat Litter at a Storage Site Was Just the Beginning," *Atlantic*, 2 November 2016, https://www.theatlantic.com. Accessed 12 June 2017.

71. And the exceptions, such as military-mandated British recycling efforts during World War II do not serve as a true analog when dealing with nuclear waste. Peter Thorsheim, *Waste into Weapons: Recycling in Britain during the Second World War* (New York: Cambridge University Press, 2015).

72. Spencer R. Weart, *The Rise of Nuclear Fear* (Cambridge: Harvard University Press, 2012), 173–74.

73. On Fukushima, see the April 2012 issue of *Environmental History*, especially Nancy Langston, "Introduction," *Environmental History* 17, no. 2 (Apr. 2012): 217–18; Sara B. Pritchard, "An Envirotechnical Disaster: Nature, Technology, and Politics at Fukushima," *Environmental History* 17, no. 2 (Apr. 2012): 219–43; Jacob Darwin Hamblin, "Fukushima and the Motifs of Nuclear History," *Environmental History* 17, no. 2 (Apr. 2012): 285–99.

CONCLUSION

1. In *Nature's Economy*, Worster cited the "father" of the atomic bomb, J. Robert Oppenheimer, as quoting the *Bhagavad-Gita* at that moment, "I am become Death, the shatterer of worlds." Donald Worster, *Nature's Economy* (San Francisco: Sierra Club Books, 1977), 339–40.

2. Chapter 6, "Ecology and the Atomic Age," in Joel B. Hagen, *An Entangled Bank: The Origins of Ecosystem Ecology* (New Brunswick: Rutgers University Press, 1992), 100–121. Also chapter 10, "Ecosystems," in Angela N. H. Creager, *Life Atomic: A History of Radioisotopes in Science and Medicine* (Chicago: University of Chicago Press, 2013).

3. On the environmental records of US leaders, especially Presidents Truman and Eisenhower, see Byron W. Daynes and Glen Sussman, *White House Politics and the Environment: Franklin D. Roosevelt to George W. Bush* (College Station: Texas A&M University Press, 2010); Riley E. Dunlap and Michael Patrick Allen, "Partisan Differences on Environmental Issues: A Congressional Roll-Call Analysis," *The Western Political Quarterly* 29, no. 3 (Sep. 1976); Elmo Richardson, *Dams, Parks, and Politics: Resource Development and Preservation in the Truman-Eisenhower Era* (Lexington: The University Press of Kentucky, 1973); Roderick Nash, review of *Dams, Parks, and Politics: Resource Development and Preservation in the Truman-Eisenhower Era* by Elmo Richardson. *American Historical Review* 80, no. 2 (Apr. 1975); Susan Hunter and Victoria Noonan, "Energy, Environment, and the Presidential Agenda," in *The Presidency Reconsidered*, ed. Richard W. Waterman (Itasca, F. E. Peacock Publishers, Inc., 1993); *The Environmental Presidency*, Dennis L. Soden, ed. (Albany: State University of New York Press,

1999); *The Environmental Legacy of Harry Truman*, Karl Boyd Brooks, ed. (Kirksville: Truman State University Press, 2009).

4. A similar process occurred in the divergence of ecology and agricultural science, where the former discipline focused on theoretical and at times conservationist thinking while the latter turned to increasing yields above other factors. Mark D. Hersey, "'What We Need is a Crop Ecologist': Ecology and Agricultural Science in Progressive-Era America," *Agricultural History* 84, no. 3 (Summer 2011), 297–321.

5. Donald Worster, "The Ecology of Order and Chaos," *Environmental History Review* 14, nos. 1–2 (Spring–Summer 1990): 1.

6. Hal K. Rothman, *The Greening of a Nation? Environmentalism in the United States Since 1945* (Fort Worth: Harcourt Brace & Company, 1998), xi.

7. Raymond Williams, *Keywords: A Vocabulary of Culture and Society* (New York: Oxford University Press, 1976, 1983), 111.

8. Michael Shellenberger and Ted Nordhaus have criticized this aspect of the environmentalism movement and called for a reexamination of what it means to be an environmentalist in their article, "The death of environmentalism." They contend, "The environmental community's narrow definition of its self-interest leads to a kind of policy literalism that undermines its power." Michael Shellenberger and Ted Nordhaus, "The Death of Environmentalism: Global Warming Politics in a Post-Environmental World." Jun 16, 2010, http://www.thebreakthrough.org. Accessed 3 June 2016.

9. He further argued that humans, being "in both kind and degree a power of a higher order than any of the other forms of animated life," could restore these "disturbed harmonies [along with] the material improvement of waste and exhausted regions." George P. Marsh, *Man and Nature; or Physical Geography as Modified by Human Action* (New York: Charles Scribner, 1864), iii.

10. On efforts to save the American Bison, particularly by the American Bison Society, see chapter 6, "The Returns of the Bison," in Andrew C. Isenberg, *The Destruction of the Bison: An Environmental History, 1750–1920* (New York: Cambridge University Press, 2000). On Progressive-Era conservation, though it is more of a work on politics at the time than environmental understandings, see Samuel P. Hays, *Conservation and the Gospel of Efficiency: The Progressive Conservation Movement, 1890–1920* (Pittsburgh: University of Pittsburgh Press, 1959, 1999). And many workers knew for years in their bodies the importance of the environment and its health to the health of their own bodies. Linda Nash, *Inescapable Ecologies: A History of Environment, Disease, and Knowledge* (Berkeley: University of California Press, 2006). Many other works, besides the ones listed below, also consider the beginning of environmentalism. For example, Adam Rome, *The Bulldozer in the Countryside: Suburban Sprawl and the Rise of American Environmentalism* (New York: Cambridge University Press, 2001); Paul S. Sutter, *Driven Wild: How the Fight against Automobiles Launched the Modern Wilderness Movement* (Seattle: University of Washington Press, 2002); Robert W. Righter, *The Battle over Hetch Hetchy: Americas Most Controversial Dam and the Birth of Modern*

Environmentalism (New York: Oxford University Press, 2005); Neil M. Maher, *Nature's New Deal: The Civilian Conservation Corps and the Roots of the American Environmental Movement* (New York: Oxford University Press, 2007); Donald Worster, *A Passion for Nature: The Life of John Muir* (New York: Oxford University Press, 2008); Karl Boyd Brooks, *Before Earth Day: The Origins of American Environmental Law, 1945–1970* (Lawrence: University Press of Kansas, 2009).

11. The book posthumously appeared a year after his death. Aldo Leopold, *A Sand County Almanac, and Sketches Here and There* (New York: Oxford University Press, 1949, 1987).

12. On scholarship questioning whether *Silent Spring* was actually a more traditionalist or conservative text, see David K. Hecht, "Rachel Carson and the Rhetoric of Revolution," *Environmental History* 24, no. 3 (July 2019): 561–82.

13. Al Gore wrote in the introduction to the 1994 reprint of the book, "*Silent Spring* came as a cry in the wilderness, a deeply felt, thoroughly researched, and brilliantly written book that changed the course of history. Without this book, the environmental movement might have been long delayed or never have developed at all." Al Gore, introduction, in Rachel Carson, *Silent Spring* (Boston: Houghton Mifflin Company, 1962, 1994), xv. Historian Ted Steinberg has explained that Carson changed "the terms of the debate over environmental reform" and "helped to transform *ecology* into the rallying cry of the environmental movement." Ted Steinberg, *Down to Earth: Nature's Role in American History* (New York: Oxford University Press, 2002), 246–47.

14. Adam Rome, *The Genius of Earth Day: How a 1970 Teach-In Unexpectedly Made the First Green Generation* (New York: Hill and Wang, 2014).

15. Otis L. Graham, Jr., *Presidents and the American Environment* (Lawrence: University Press of Kansas, 2015), 366. Graham had little positive to say about either Truman or Eisenhower, summarizing, "The fifteen years after FDR's death had been a discouraging time for the conservation movement," as membership of national organizations dropped and few new ones formed, and the two presidents only sporadically did anything to protect the natural world, almost entirely ignoring air and water pollution and agricultural pesticide hazards. See pp. 153–86.

16. Though the ecomodernism of their Breakthrough Institute is controversial, this sentiment is similar to the assertion of environmental activists Michael Shellenberger and Ted Nordhaus, who argued in 2004, "The entire landscape in which politics plays out has changed radically in the last 30 years, yet the environmental movement acts as though proposals based on 'sound science' will be sufficient to overcome ideological and industry opposition." Shellenberger and Nordhaus, "The Death of Environmentalism."

17. Andrew J. Bacevich, *The New American Militarism: How Americans Are Seduced by War* (New York: Oxford University Press, 2005), 1.

18. Lawrence H. Keeley, *War before Civilization* (New York: Oxford University Press, 1996), 165.

19. When the expected Iranian civilian fatalities were upped twenty-fold to two million, 59 percent still supported the bomb use. YouGov carried out the 2016 poll, which had 840 respondents. Scott D. Sagan and Benjamin A. Valentino, "Would the U.S. Drop the Bomb Again?" *Wall Street Journal*, 19 May 2016, http://www.wsj.comon. Accessed on 2 June 2016.

20. Etienne Benson, *Surroundings: A History of Environments and Environmentalisms* (University of Chicago Press, 2020), 13.

21. Scruton, who has since been knighted, is a former professor of aesthetics at Birkbeck College, London, and wrote a text on the issue titled *How to Think Seriously About the Planet: The Case for Environmental Conservatism*. Paul Basken, "Scholar of Aesthetics Hopes to Show Conservatives Their Inner Environmentalist," *Chronicle of Higher Education*, 5 June 2012, http://www.chronicle.com. Accessed on 19 June 2017.

22. Dwight D. Eisenhower Presidential Library, Abilene, Kansas (hereafter DDEL), White House Central Files, Official File, Box 213, Folder 8, Commencement Address by Thomas E. Murray at Seattle University, Seattle, WA, 29 May 1953, 3–4.

23. Murray of course overstated his case. Even if so-called natural disasters do possess a significant cultural component, human capabilities fall well short of what Murray described. Ted Steinberg, *Acts of God: The Unnatural History of Natural Disaster in America* (New York: Oxford University Press, 2000).

24. Matthew Morse Booker, *Down by the Bay: San Francisco's History between the Tides* (Berkeley: University of California Press, 2013), 9.

Bibliography

PRIMARY SOURCES
Archival Collections
Defense Technical Information Center, Fort Belvoir, Virginia (DTIC)
Dwight D. Eisenhower Presidential Library, Abilene, Kansas (DDEL)
   Dwight D. Eisenhower Papers as President, Ann Whitman File
      Administrative Series
      Cabinet Series
      Campaign Series
      Diary Series
      Press Conference Series
      Speech Series
   John Stewart Bragdon Records, Miscellaneous File
   John Foster Dulles Papers, General Correspondence and Memoranda Series
   John A. McCone Papers
   White House Central Files
      Confidential Files
      General File
      Official File
      Official File, Cross Reference Sheets
   White House Office
      Office of the Special Assistant, OCB Series
         Administrative Subseries
         Subject Subseries
      Office of the Staff Secretary
         Records, Subject Series, Alphabetical Subseries
Harry S. Truman Presidential Library, Independence, Missouri (HSTL)
   Atomic Bomb Collection

Dean G. Acheson Papers
Papers of Clark Clifford, Subject File
Papers of George M. Elsey
Papers of Harry S. Truman
  Official File
  President's Secretary's Files
White House Central Files, Confidential Files
The National Archives at College Park, MD (NACP)
  RG 326 Records of the Atomic Energy Commission
    Entry A1 19, Minutes of the Meetings of the AEC
    Entry 73, Division of Biology and Medicine, Records Relating to Fallout
      Studies, 1953–6
    Entry 81 A, Commissioner Harold S. Vance, Correspondence, 1955–1959
  RG 374 Records of the Defense Threat Reduction Agency

PUBLISHED PRIMARY SOURCES

Alsop, Joseph, and Stewart Alsop. "How Red A-Blast Was Detected." *Washington Post*, 31 December 1950.

"Atoms Vital to Agriculture." *Science News Letter* 69, no. 3 (21 January 1956): 35.

Bible (New International Version translation). Biblica, 1978.

Carson, Rachel. *Silent Spring*. Boston: Houghton Mifflin Company, 1962, 1994.

Commoner, Barry. "Conservation of the Water Resource: The Responsibility of the Scientist," *Journal (Water Pollution Control Federation)* 37, no. 1 (Jan. 1965): 60–70.

———. "The Fallout Problem," *Science* 127, no. 3305 (May 2, 1958): 1023–26.

Federal Civil Defense Administration, "Bert the Turtle Says Duck and Cover." US Government Printing Office, 1951.

Gamarekian, Edward. "HEW Acts to End A-Contamination of Rivers," *Washington Post and Times Herald*, 15 July 1959.

Hersey, John. *Hiroshima*. New York: Bantam Books, Inc., 1946, 1981.

Kistiakowsky, George B. *A Scientist at the White House: The Private Diary of President Eisenhower's Special Assistant for Science and Technology*. Cambridge: Harvard University Press, 1976.

"Major Describes Move," *New York Times*, 8 February 1968, 14.

Musial, Joe. *Learn How Dagwood Splits the Atom!* King Features Syndicate, Inc., 1949.

National Academy of Sciences—National Research Council, "Radioactive Waste Disposal into Atlantic and Gulf Coast Waters," Publication 655, 1959.

Reports of the Atomic Energy Commission to Congress. Washington, DC: United States Government Printing Office.

———. *Fifth Semiannual Report, January 1949*.

———. *Sixth Semiannual Report, July 1949*.

————. *Seventh Semiannual Report, January 1950.*

————. *Eighth Semiannual Report, July 1950.*

————. *Ninth Semiannual Report, January 1951.*

————. *Tenth Semiannual Report, July 1951.*

————. *Eleventh Semiannual Report, January 1952.*

————. *Twelfth Semiannual Report, July 1952.*

————. *Thirteenth Semiannual, January 1953.*

————. *Fourteenth Semiannual, July 1953.*

————. *Fifteenth Semiannual Report, January 1954.*

————. *Sixteenth Semiannual Report, July 1954.*

————. *Seventeenth Semiannual Report, January 1955.*

————. *Eighteenth Semiannual Report, July 1955.*

————. *Nineteenth Semiannual Report, January 1956.*

————. *Twentieth Semiannual Report, July 1956.*

————. *Twenty-First Semiannual Report, January 1957.*

————. *Twenty-second Semiannual Report, July 1957.*

————. *Twenty-third Semiannual Report, January 1958.*

————. *Twenty-fourth Semiannual Report, July 1958.*

————. *Twenty-fifth Semiannual Report, January 1959.*

————. *Annual Report for 1959, January 1960.*

————. *Annual Report for 1960, January 1961.*

"The Unlucky Dragon," *Washington Post and Times Herald*, 19 March 1954, 28.

Time L:5 (4 August 1947).

US Bureau of Labor Statistics, "100 Years of U.S. Consumer Spending: Data for the Nation, New York City, and Boston." Report 991, May 2006.

VIDEO

Rizzo, Anthony, dir. *Duck and Cover.* 1952. Archer Productions.

Stone, Robert, dir. *Pandora's Promise,* 2013. Robert Stone Productions, Vulcan Productions.

WEBSITES

Atlantic, https://www.theatlantic.com.

Breakthrough Institute, http://www.thebreakthrough.org.

Chronicle of Higher Education, http://www.chronicle.com.

Commentary, http://www.commentarymagazine.com.

Dwight D. Eisenhower Presidential Library, Museum and Boyhood Home, http://www.eisenhower.archives.gov.

Johnson, Paige, "Atomic Gardening: An Online History," http://www.atomicgardening.com.

Internet Archive, http://archive.org.

Slate, http://www.slate.com.

Nukewatch, http://nukewatchinfo.org.

Popular Mechanics http://www.popularmechanics.com.

US Bureau of Labor Statistics, http://www.bls.gov.

Wall Street Journal, http://www.wsj.com/.

YouTube, http://www.youtube.com.

SECONDARY SOURCES

Adams, Melvin R. *Atomic Geography: A Personal History of the Hanford Nuclear Reservation*. Pullman: Washington State University Press, 2016.

Ahlberg, Kristin L. *Transplanting the Great Society: Lyndon Johnson and Food for Peace*. Columbia: University of Missouri Press, 2008.

Anderson, J. L. *Industrializing the Corn Belt: Agriculture, Technology, and Environment, 1945–1972*. DeKalb: Northern Illinois University Press, 2009.

Bacevich, Andrew J. *The New American Militarism: How Americans Are Seduced by War*. New York: Oxford University Press, 2005.

Ball, Howard. *Justice Downwind: America's Atomic Testing Program in the 1950s*. New York: Oxford University Press, 1986.

Basken, Paul. "Scholar of Aesthetics Hopes to Show Conservatives Their Inner Environmentalist." *Chronicle of Higher Education*, 5 June 2012.

Barker-Devine, Jenny. "'Mightier Than Missiles': The Rhetoric of Civil Defense for Rural American Families, 1950–1970." *Agricultural History* 80:4 (Autumn, 2006): 415–35.

Barth, Kai-Henrik. "Detecting the Cold War: Seismology and Nuclear Weapons Testing, 1945–1970." PhD diss., University of Minnesota, 2000.

Balogh, Brian. *Chain Reaction: Expert Debate and Public Participation in American Commercial Nuclear Power, 1945–1975*. Cambridge: Cambridge University Press, 1991.

Benson, Etienne. *Surroundings: A History of Environments and Environmentalisms*. University of Chicago Press, 2020.

Bess, Michael. *The Light-Green Society: Ecology and Technological Modernity in France, 1960–2000*. Chicago: The University of Chicago Press, 2003.

Bijker, Wiebe E. *Of Bicycles, Bakelites, and Bulbs: Toward a Theory of Sociotechnical Change*. Cambridge: The MIT Press, 1995.

Bird, Kai, and Martin J. Sherwin. *American Prometheus: The Triumph and Tragedy of J. Robert Oppenheimer*. New York: Alfred A. Knopf, 2005.

Black, Megan. *The Global Interior: Mineral Frontiers and American Power*. Harvard University Press, 2018.

Blitz, Matt. "When Kodak Accidentally Discovered A-Bomb Testing," *Popular Mechanics*. 20 June 2016.

Booker, Matthew Morse. *Down by the Bay: San Francisco's History between the Tides*.

Berkeley: University of California Press, 2013.

Bordsen, John. "NC 12 Is Secret to the Outer Banks Popularity," *News and Observer* (Raleigh, NC), 6 May 2016.

Boyer, Paul. *By the Bomb's Early Light: American Thought and Culture at the Dawn of the Atomic Age.* New York: Pantheon Books, 1985.

_____. *Fallout: A Historian Reflects on America's Half-Century Encounter with Nuclear Weapons.* Columbus: Ohio State University Press, 1998.

Boynton, Alex. "Formulating an Anti-Environmental Opposition: Neoconservative Intellectuals during the Environmental Decade," *The Sixties: A Journal of History, Politics and Culture* 8 no. 1 (Jun. 2015), 1–26.

Brady, Lisa M. *War Upon the Land: Military Strategy and the Transformation of Southern Landscapes During the American Civil War.* Athens: University of Georgia Press, 2012.

_____. "The Wilderness of War: Nature and Strategy in the American Civil War." *Environmental History* 10:3 (Jul. 2005): 421–47.

Brooks, Karl Boyd. *Before Earth Day: The Origins of American Environmental Law, 1945–1970.* Lawrence: University Press of Kansas, 2009.

_____, ed. *The Environmental Legacy of Harry Truman.* Kirksville: Truman State University Press, 2009.

Brown, Kate. *Plutopia: Nuclear Families, Atomic Cities, and the Great Soviet and American Plutonium Disasters.* New York: Oxford University Press, 2012.

Carter, Donald Alan. "Eisenhower versus the Generals." *The Journal of Military History* 71:4 (Oct. 2007): 1169–99.

Cloud, John. "Introduction: Special Guest-Edited Issue on the Earth Sciences in the Cold War," *Social Studies of Science* 33:5 (Oct. 2003), 629–33.

Coclanis, Peter A. "Field Notes: Agricultural History's New Plot." *Journal of Interdisciplinary History* 50, no. 2 (Autumn, 2019), 187–212.

Collins, Harry, and Trevor Pinch, *The Golem: What You Should Know about Science, Second Edition.* New York: Cambridge University Press, 1993, 1998.

_____. *The Golem at Large: What You Should Know about Technology.* New York: Cambridge University Press, 1998.

Creager, Angela N. H. *Life Atomic: A History of Radioisotopes in Science and Medicine.* Chicago: University of Chicago Press, 2013.

Cronon, William, ed. *Uncommon Ground: Rethinking the Human Place in Nature* (New York: W. W. Norton & Company, 1995, 1996)

Cullather, Nick. *The Hungry World: America's Cold War Battle against Poverty in Asia.* Cambridge: Harvard University Press, 2010.

Curry, Helen Anne. "Atoms in Agriculture." *Historical Studies in the Natural Sciences* 46, no. 2 (April 2016): 119–53.

Dalton, Russell J., Paula Garb, Nicholas P. Lovrich, John C. Pierce, John M. Whiteley, *Critical Masses: Citizens, Nuclear Weapons Production, and Environmental*

Destruction in the United States and Russia. Cambridge: Massachusetts Institute of Technology Press, 1999.

Danbom, David B. *Born in the Country: A History of Rural America.* Baltimore: The Johns Hopkins University Press, 1995.

Davis, Mike. *Late Victorian Holocausts: El Niño Famines and the Making of the Third World.* New York: Verso, 2002.

Daynes, Byron W., and Glen Sussman. *White House Politics and the Environment: Franklin D. Roosevelt to George W. Bush.* College Station: Texas A&M University Press, 2010.

DeLoughrey, Elizabeth. "Radiation Ecologies and the Wars of Light." *Modern Fiction Studies* 55, no. 3 (Fall 2009): 468–95.

Diefendorf, Jeffry M., and Kurk Dorsey, eds. *City, Country, Empire: Landscapes in Environmental History.* Pittsburgh: University of Pittsburgh Press, 2005.

Divine, Robert A. *Blowing on the Wind: The Nuclear Test Ban Debate, 1954–1960.* New York: Oxford University Press, 1978.

———. "Eisenhower, Dulles, and the Nuclear Test Ban Issue: Memorandum of a White House Conference, 24 March 1958." *Diplomatic History* 2, no. 1 (Oct. 1978): 321–30.

Dorsey, Kurk, and Mark Lytle. "Introduction." *Diplomatic History* 32, no. 4 (Sep. 2008): 517–18.

Dunlap, Riley E., and Michael Patrick Allen. "Partisan Differences on Environmental Issues: A Congressional Roll-Call Analysis." *Western Political Quarterly* 29 no. 3 (Sep. 1976): 384–97.

Egan, Michael. *Barry Commoner and the Science of Survival: The Remaking of American Environmentalism.* Boston: MIT Press, 2007.

Ellul, Jacques. *The Technological Society,* Translated by John Wilkinson. New York: Alfred A Knopf, 1964.

Evans, Sterling. "The 'Age of Agricultural Ignorance': Trends and Concerns for Agriculture into the Twenty-First Century." *Agricultural History* 93, no. 1 (Winter 2019): 4–34.

Fiege, Mark. "The Atomic Scientists, the Sense of Wonder, and the Bomb." *Environmental History* 12, no. 3 (Jul. 2007): 578–613.

———. *The Republic of Nature: An Environmental History of the United States* Seattle: University of Washington Press, 2012, 2013.

Finkbeiner, Ann. "How Do We Know Nuclear Bombs Blow Down Forests? Because We Built a Forest in Nevada and Blew It Down," *Slate,* 31 May 2013.

Fitzgerald, Deborah. *Every Farm a Factory: The Industrial Ideal in American Agriculture.* New Haven: Yale University Press, 2003.

Fleming, James Rodger. *Fixing the Sky: The Checkered History of Weather and Climate Control.* New York: Columbia University Press, 2010.

Fleming, James Rodger, and Ann Johnson, eds. *Toxic Airs: Body, Place, Planet in*

Historical Perspective. University of Pittsburgh Press, 2014.

Foxley, Curtis Frederick. "The Business of Atomic War: The Military-Industrial Complex and the American West." PhD diss., University of Oklahoma, 2020.

Fradkin, Phillip L. *Fallout: An American Nuclear Tragedy*. Tucson: University of Arizona Press, 1989.

Gilinsky, Victor. "Yucca Mountain redux." *Bulletin of the Atomic Scientists*. 5 November 2014.

Gore, Al. "Introduction," in Rachel Carson, *Silent Spring*, xv–xxvi. Boston: Houghton Mifflin Company, 1962, 1994.

Graham, Jr., Otis L. *Presidents and the American Environment*. Lawrence: University Press of Kansas, 2015.

Greene, Benjamin P. *Eisenhower, Science Advice, and the Nuclear Test-Ban Debate, 1945–1963*. Stanford: Stanford University Press, 2007.

Hacker, Barton C. *Elements of Controversy: The Atomic Energy Commission and Radiation Safety in Nuclear Weapons Testing, 1947–1974*. Berkeley: University of California Press, 1994.

Hagen, Joel B. *An Entangled Bank: The Origins of Ecosystem Ecology*. New Brunswick: Rutgers University Press, 1992.

Hamblin, Jacob Darwin. *Arming Mother Nature: The Birth of Catastrophic Environmentalism*. New York: Oxford University Press, 2013.

————. "Fukushima and the Motifs of Nuclear History." *Environmental History* 17, no. 2 (Apr. 2012): 285–99.

————. "Gods and Devils in the Details: Marine Pollution, Radioactive Waste, and an Environmental Regime circa 1972." *Diplomatic History* 32, no. 4 (Sep. 2008): 539–60.

————. "Let There Be Light . . . and Bread: The United Nations, the Developing World, and Atomic Energy's Green Revolution." *History and Technology* 25, no. 1 (Mar. 2009): 25–48.

————. *Poison in the Well: Radioactive Waste in the Oceans at the Dawn of the Nuclear Age*. New Brunswick: Rutgers University Press, 2008.

————. *The Wretched Atom: America's Global Gamble with Peaceful Nuclear Technology*. New York: Oxford University Press, 2021.

Hamilton, Shane. *Supermarket USA: Food and Power in the Cold War Farms Race*. New Haven: Yale University Press, 2018.

————. *Trucking Country: The Road to America's Wal-Mart Economy*. Princeton: Princeton University Press, 2008.

Harper, Kristine C. "Research from the Boundary Layer: Civilian Leadership, Military Funding and the Development of Numerical Weather Prediction (1946–55)." *Social Studies of Science* 33, no. 5 (Oct. 2003): 667–96.

Hart, John Fraser. *The Land That Feeds Us*. New York: W. W. Norton & Company, 1991.

Hays, Samuel P. *Conservation and the Gospel of Efficiency: The Progressive Conservation Movement, 1890–1920.* Pittsburgh: University of Pittsburgh Press, 1959, 1999.

Hecht, David K. "Rachel Carson and the Rhetoric of Revolution." *Environmental History* 24, no. 3 (Jul. 2019): 561–82.

Hecht, Gabrielle. *Being Nuclear: Africans and the Global Uranium Trade.* MIT Press, 2012.

———. *The Radiance of France: Nuclear Power and National Identity after World War II.* Cambridge: MIT Press, 1998, 2009.

Hersey, Mark D. "'What We Need Is a Crop Ecologist': Ecology and Agricultural Science in Progressive-Era America." *Agricultural History* 84, no. 3 (Summer 2011): 297–321.

Hevly, Bruce, and John M. Findlay, eds. *The Atomic West.* Seattle: University of Washington Press, 1998.

Hewlett, Richard G., and Oscar E. Anderson, Jr. *The New World, 1939/1946: Volume I of a History of the United States Atomic Energy Commission.* University Park: Pennsylvania University Press, 1962.

Hewlett, Richard G., and Francis Duncan. *Atomic Shield, 1947/1952: Volume II of A History of the United States Atomic Energy Commission.* University Park: Pennsylvania State University Press, 1969.

Hewlett, Richard G., and Jack M. Holl. *Atoms for Peace and War, 1953–1961: Eisenhower and the Atomic Energy Commission.* Berkeley: University of California Press, 1989.

Higuchi, Toshihiro. *Political Fallout: Nuclear Weapons Testing and the Making of a Global Environmental Crisis.* Stanford University Press, 2020.

Isenberg, Andrew C. *The Destruction of the Bison: An Environmental History, 1750–1920.* New York: Cambridge University Press, 2000.

Jackson, Michael Gordon. "Beyond Brinkmanship: Eisenhower, Nuclear War Fighting, and Korea, 1953–1968." *Presidential Studies Quarterly* 35, no. 1 (Mar. 2005): 52–75.

Jacobs, Robert A. *The Dragon's Tail: American's Face the Atomic Age.* Amherst and Boston: University of Massachusetts Press, 2010.

Jacobson, Harold K., and Eric Stein, *Diplomats, Scientists and Politicians: The United States and the Nuclear Test Ban Negotiations.* Ann Arbor, University of Michigan Press, 1966.

Jessee, Emery Jerry. "Radiation Ecologies: Bombs, Bodies, and Environment during the Atmospheric Nuclear Weapons Testing Period, 1942–1965." PhD diss., Montana State University, 2013.

Jenks, Andrew. "Model City USA: The Environmental Cost of Victory in World War II and the Cold War." *Environmental History* 12, no. 3 (Jul. 2007): 552–77.

Joes, Anthony James. "Eisenhower Revisionism: The Tide Comes In." *Presidential Studies Quarterly* 15, no. 3 (Summer, 1985): 561–71.

Johnston, Barbara Rose, and Holly M. Barker. *The Rongelap Report: Consequential Damages of Nuclear War*. Walnut Creek: Left Coast Press, 2008.

Jørgensen, Dolly. "Not by Human Hands: Five Technological Tenets for Environmental History in the Anthropocene." *Environment and History* 20, no. 4 (Nov. 2014): 479–89.

Jørgensen, Dolly, Finn Arne Jørgensen, and Sara B. Pritchard, eds., *New Natures: Joining Environmental History with Science and Technology Studies*. Pittsburgh: University of Pittsburgh Press, 2013.

Keeley, Lawrence H. *War before Civilization*. New York: Oxford University Press, 1996.

Kirsch, Scott. "Ecologists and the Experimental Landscape: The Nature of Science at the US Department of Energy's Savannah River Site." *Cultural Geographies* 14, no. 4 (Oct. 2007): 485–510.

————. *Proving Grounds: Project Plowshare and the Unrealized Dream of Nuclear Earthmoving*. New Brunswick: Rutgers University Press, 2005.

Kohler, Robert E. *Landscapes and Labscapes: Exploring the Lab-Field Border in Biology*. Chicago: University of Chicago Press, 2002.

Kolhoff, Dean W. *Amchitka and the Bomb: Nuclear Testing in Alaska*. Seattle: University of Washington Press, 2002.

Kolata, Gina. "In Good Health? Thank Your 100 Trillion Bacteria," *New York Times*, 13 June 2012.

Kuhn, Thomas S. *The Structure of Scientific Revolutions*. Chicago: University of Chicago Press, 1962, 2012.

Kuletz, Valerie. *The Tainted Desert: Environmental and Social Ruin in the American West*. New York: Routledge, 1998.

Laakkonen, Simo, Viktor Pál, and Richard Tucker, "The Cold War and Environmental History: Complementary Fields." *Cold War History* 16, no. 4 (Fall 2016): 377–94.

Lachmund, Jens. "Exploring the City of Rubble: Botanical Fieldwork in Bombed Cities in Germany after World War II." *Osiris* 18 (2003): 234–54.

Langston, Nancy. "Introduction." *Environmental History* 17, no. 2 (Apr. 2012): 217–18.

————, *Toxic Bodies: Hormone Disruptors and the Legacy of DES*. New Haven: Yale University Press, 2010.

Lavine, Matthew. *The First Atomic Age: Scientists, Radiations, and the American Public, 1895–1945*. New York: Palgrave Macmillan, 2013.

Leopold, Aldo. *A Sand County Almanac, and Sketches Here and There*. New York: Oxford University Press, 1949, 1987.

Lipartito, Kenneth. "Picturephone and the Information Age: The Social Meaning of Failure." *Technology and Culture* 44, no. 1 (Jan. 2003): 50–81.

Lutts, Ralph H. "Chemical Fallout: Rachel Carson's Silent Spring, Radioactive

Fallout, and the Environmental Movement." *Environmental Review: ER* 9, no. 3 (Autumn 1985): 210–25.

Maher, Neil M. *Nature's New Deal: The Civilian Conservation Corps and the Roots of the American Environmental Movement.* New York: Oxford University Press, 2007.

Martin, Laura J. "*Proving Grounds: Ecological Fieldwork in the Pacific and the Materialization of Ecosystems.*" *Environmental History* 23, no. 3 (Jul. 2018): 567–92.

Macfarlane, Allison. "Underlying Yucca Mountain: The Interplay of Geology and Policy in Nuclear Waste Disposal." *Social Studies of Science* 33, no. 5 (Oct. 2003): 783–807.

MacLeod, Roy M. *Science and the Pacific War: Science and Survival in the Pacific, 1939–1945.* Hingham: Kluwer Academic Publishers, 2000.

Maier, Charles S. "Introduction: Science, Politics, and Defense in the Eisenhower Era," in *A Scientist at the White House: The Private Diary of President Eisenhower's Special Assistant for Science and Technology,* by George B. Kistiakowsky. Cambridge: Harvard University Press, 1976.

Makhijani, Arjun, Howard Hu, and Katherine Yih, eds. *Nuclear Wastelands: A Global Guide to Nuclear Weapons Production and Its Health and Environmental Effects.* Cambridge: MIT Press, 1995.

Malin, Stephanie A. *The Price of Nuclear Power: Uranium Communities and Environmental Justice.* New Brunswick: Rutgers University Press, 2015.

Marsh, George P. *Man and Nature; or Physical Geography as Modified by Human Action.* New York: Charles Scribner, 1864.

Martini, Edwin, ed., *Proving Grounds: Weapons Testing, Militarized Landscapes, and the Environmental Impact of American Empire.* Seattle: University of Washington Press, 2015.

Masco, Joseph. *The Nuclear Borderlands: The Manhattan Project in Post–Cold War New Mexico.* Princeton University Press, 2006.

Matthews, Jr., Melvin E. *Duck and Cover: Civil Defense Images in Film and Television from the Cold War to 9/11.* Jefferson: McFarland & Company, Inc., 2012.

McAuliffe, Mary S. "Eisenhower, the President," *The Journal of American History* 68, no. 3 (Dec. 1981): 625–32.

McBride, James Hubert. *The Test Ban Treaty: Military, Technological, and Political Implications.* Chicago: Henry Regnery Company, 1967.

McNeill, J. R. *Something New under the Sun: An Environmental History of the Twentieth Century World.* New York: W. W. Norton & Company, 2001.

———. "Woods and Warfare in World History." *Environmental History* 9, no. 3 (Jul. 2004): 388–410.

McNeill, J. R. and Corinna R. Unger, eds. *Environmental Histories of the Cold War,* New York: Cambridge University Press, 2010.

Melosi, Martin V. *Atomic Age America.* Boston: Pearson, 2012.

Meyers, Keith Andrew. "In the Shadow of the Mushroom Cloud: Nuclear Testing,

Radioactive Fallout, and Damage to U.S. Agriculture, 1945–1970." *Journal of Economic History* 79, no. 1 (Mar. 2019): 244–74.

———. "Investigating the Economic Consequences of Atmospheric Nuclear Testing." Dissertation, University of Arizona, 2018.

Miller, Richard L. *Under the Cloud: The Decades of Nuclear Testing*. New York: The Free Press, 1986.

Moore, Alan, and Dave Gibbons. *Watchmen*. New York: DC Comics, 1986, 2005.

Mumford, Lewis. *Technics and Civilization*. New York: Harcourt, Brace and Company, 1934.

Nash, Linda. *Inescapable Ecologies: A History of Environment, Disease, and Knowledge*. Berkeley: University of California Press, 2006.

Nash, Roderick. Review of *Dams, Parks & Politics: Resource Development and Preservation in the Truman-Eisenhower Era* by Elmo Richardson. *American Historical Review* 80, no. 2 (Apr. 1975): 538–39.

Neufeld, Jacob, George M. Watson Jr., and David Chenoweth, eds. *Technology and the Air Force: A Retrospective Assessment*. Washington, DC: Air Force History and Museums Program, 1997.

The Nevada Test Site: A Guide to America's Nuclear Proving Ground. Culver City: The Center for Land Use Interpretation, 1996.

Novick, Sheldon. *The Carless Atom*. Boston: Houghlin Mifflin, 1969.

———. "The Menace of the Peaceful Atom," *Commentary*, 1 December 1968, 33–40.

Oliver, Kendrick. *Kennedy, Macmillan and the Nuclear Test-Ban Debate, 1961–63*. New York: St. Martin's Press, Inc., 1998.

O'Lear, Shannon. *Environmental Politics: Scale and Power*. New York: Cambridge University Press, 2010.

O'Neill, Dan. *The Firecracker Boys*. New York: St. Martin's Press, 1994.

Oreskes, Naomi, and Erik M. Conway. *Merchants of Doubt: How a Handful of Scientists Obscured the Truth on Issues from Tobacco Smoke to Global Warming*. New York: Bloomsbury Press, 2010.

Pacey, Arnold. *Meaning in Technology*. Cambridge: MIT Press, 1999.

Pearson, Chris. *Scarred Landscapes: War and Nature in Vichy France*. New York: Palgrave Macmillan, 2008.

Pearson, Chris, Peter Coates, and Tim Cole, eds., *Militarized Landscapes: From Gettysburg to Salisbury Plain*. New York: Continuum, 2010.

Perkins, John H. *Geopolitics and the Green Revolution: Wheat, Genes, and the Cold War*. New York: Oxford University Press, 1997.

Perrow, Charles. *Normal Accidents: Living with High-Risk Technologies*. New York: Basic Books, Inc., 1984.

Petersen, Melody. "As Beef Cattle Become Behemoths, Who Are Animal Scientists Serving?" *Chronicle of Higher Education*, 15 April 2012.

Pharo, Per Fredrik Ilsaas. "A Precondition for Peace: Transparency and the Test-Ban Negotiations, 1958–1963." *The International History Review* 22, no. 3 (Sep. 2000): 557–82.

Pilcher, Jeffrey M., ed. *The Oxford Handbook of Food History.* Oxford University Press, 2012.

Pritchard, Sara B. "An Envirotechnical Disaster: Nature, Technology, and Politics at Fukushima." *Environmental History* 17, no. 2 (Apr. 2012): 219–43.

Raby, Megan. *American Tropics: The Caribbean Roots of Biodiversity Science.* Chapel Hill: University of North Carolina Press, 2017.

Radkau, Joachim. *Nature and Power: A Global History of the Environment,* trans. Thomas Dunlap. New York: Cambridge University Press, 2002, 2008.

Reuss, Martin, and Stephen H. Cutcliffe, eds. *The Illusory Boundary: Environment and Technology in History.* Charlottesville: University of Virginia Press, 2010.

Righter, Robert W. *The Battle over Hetch Hetchy: Americas Most Controversial Dam and the Birth of Modern Environmentalism.* New York: Oxford University Press, 2005.

Richardson, Elmo. *Dams, Parks & Politics: Resource Development and Preservation in the Truman-Eisenhower Era.* Lexington: The University Press of Kentucky, 1973.

———. *The Presidency of Dwight D. Eisenhower.* Lawrence: The Regents Press of Kansas, 1979.

Rome, Adam. *The Bulldozer in the Countryside: Suburban Sprawl and the Rise of American Environmentalism.* New York: Cambridge University Press, 2001.

———. *The Genius of Earth Day: How a 1970 Teach-In Unexpectedly Made the First Green Generation.* New York: Hill and Wang, 2014.

Rosenthal, Debra. *At the Heart of the Bomb: The Dangerous Allure of Weapons Work.* Reading: Addison-Wesley Publishing, Inc., 1990.

Rothman, Hal K. *The Greening of a Nation? Environmentalism in the United States Since 1945.* Fort Worth: Harcourt Brace & Company, 1998.

Edmund Russell. *War and Nature: Fighting Humans and Insects with Chemicals from World War I to Silent Spring.* New York: Cambridge University Press, 2001.

Sagan, Scott D., and Benjamin A. Valentino. "Would the U.S. Drop the Bomb Again?" *Wall Street Journal,* 19 May 2016.

Schapsmeier, Edward L., and Frederick H. Schapsmeier. "Eisenhower and Agricultural Reform: Ike's Farm Policy Legacy Appraised." *American Journal of Economics and Sociology* 51, no. 2 (Apr. 1992): 147–59.

Schrepfer, Susan R., and Philip Scranton, eds. *Industrializing Organisms: Introducing Evolutionary History.* New York: Routledge, 2004.

Schwelb, Egon. "The Nuclear Test Ban Treaty and International Law." *American Journal of International Law* 58, no. 3 (Jul. 1964): 642–70.

Seaborg, Glenn T. *Kennedy, Khrushchev, and the Test Ban.* Berkeley: University of California Press, 1981.

Selcer, Perrin. *The Postwar Origins of the Global Environment: How the United Nations Built Spaceship Earth.* Columbia University Press, 2018.

Shellenberger, Michael, and Ted Nordhaus, "The Death of Environmentalism: Global Warming Politics in a Post–Environmental World." Paper presented at the October 2004 meeting of the Environmental Grantmakers Association.

Sherry, Michael. *The Rise of American Air Power: The Creation of Armageddon.* New Haven: Yale University Press, 1987.

Smith-Norris, Martha. "The Eisenhower Administration and the Nuclear Test Ban Talks, 1958–1960: Another Challenge to Revisionism." *Diplomatic History* 27, no. 4 (Aug. 2003): 503–41.

Soden, Dennis L., ed. *The Environmental Presidency.* Albany: State University of New York Press, 1999.

Steinberg, Ted. *Acts of God: The Unnatural History of Natural Disaster in America.* New York: Oxford University Press, 2000.

————. *Down to Earth: Nature's Role in American History.* New York: Oxford University Press, 2002.

Stenehjem, Michele A. "Pathways of Radioactive Contamination: Beginning the History, Public Enquiry, and Policy Study of the Hanford Nuclear Reservation." *Environmental Review* 13, nos. 3/4 (Autumn–Winter, 1989): 94–112.

Sutter, Paul S. *Driven Wild: How the Fight against Automobiles Launched the Modern Wilderness Movement.* Seattle: University of Washington Press, 2002.

————. "The World with Us: The State of American Environmental History," *Journal of American History* 100, no. 1 (Jun. 2013), 94–119.

Szasz, Ferenc M. *The Day the Sun Rose Twice: The Story of the Trinity Nuclear Explosion, July 16, 1945.* Albuquerque: University of New Mexico Press, 1985.

Thee, Marek. "The Pursuit of a Comprehensive Nuclear Test Ban." *Journal of Peace Research* 25, no. 1 (Mar. 1988): 5–15.

Thomas, Julia Adeney. "History and Biology in the Anthropocene: Problems of Scale, Problems of Value," *American Historical Review* 119, no. 5 (Dec. 2014): 1587–1607.

————. *Reconfiguring Modernity: Concepts of Nature in Japanese Political Ideology.* Berkeley: University of California Press, 2001.

Thorsheim, Peter. *Waste into Weapons: Recycling in Britain during the Second World War.* New York: Cambridge University Press, 2015.

Tucker, Richard P., and Edmund Russell, eds. *Natural Enemy, Natural Ally: Toward an Environmental History of War.* Corvallis: Oregon State University Press, 2004.

Tsutsui, William M. *Godzilla on My Mind: Fifty Years of the King of Monsters.* New York: Palgrave Macmillan, 2004.

————. "Looking Straight at *Them!* Understanding the Big Bug Movies of the 1950s." *Environmental History* 12, no. 2 (Apr. 2007): 237–53.

Vail, David. *Chemical Lands: Pesticides, Aerial Spraying, and Health in North America's Grasslands since 1945.* Tuscaloosa: University of Alabama Press, 2018.

Vetter, Jeremy. *Field Life: Science in the American West during the Railroad Era*. Pittsburgh: University Pittsburgh Press, 2016.

Walker, J. Samuel, ed. *Nuclear Energy and the Legacy of Harry S. Truman*. Kirksville: Truman State University Press, 2016.

———. *Prompt and Utter Destruction: Truman and the Use of Atomic Bombs against Japan*. Chapel Hill: University of North Carolina Press, 1997, 2004.

Waterman, Richard W., ed. *The Presidency Reconsidered*. Itasca: F. E. Peacock Publishers, Inc., 1993.

Weart, Spencer R. *Scientists in Power*. Cambridge: Harvard University Press, 1979.

———. *The Rise of Nuclear Fear*. Cambridge: Harvard University Press, 2012.

White, Richard. *The Organic Machine: The Remaking of the Columbia River*. New York: Hill and Wang, 1995.

Williams, Raymond. *Keywords: A Vocabulary of Culture and Society*. New York: Oxford University Press, 1976, 1983.

Williams, Rosalind. "Classics Revisited: Lewis Mumford's *Technics and Civilization*." *Technology and Culture* 43, no. 1 (Jan. 2002): 139–49.

Winner, Langdon. *Autonomous Technology: Technics-Out-of-Control as a Theme in Political Thought*. Cambridge: MIT Press, 1977.

Wolfe, Audra J. *Freedom's Laboratory: The Cold War Struggle for the Soul of Science*. Baltimore: Johns Hopkins University Press, 2018.

Worster, Donald. *Dust Bowl: The Southern Plains in the 1930s*. New York: Oxford University Press, 1979.

———. *Nature's Economy*. San Francisco: Sierra Club Books, 1977.

———. "The Ecology of Order and Chaos." *Environmental History Review* 14, nos. 1–2 (Spring–Summer, 1990): 1–18.

———. *A Passion for Nature: The Life of John Muir*. New York: Oxford University Press, 2008.

———. *Shrinking the Earth: The Rise & Decline of American Abundance*. New York: Oxford University Press, 2016.

———. "Transformations of the Earth: Toward an Agroecological Perspective in History." *The Journal of American History* 76, no. 4 (Mar. 1990): 1087–1106.

Wright, Angus. *The Death of Ramón González: The Modern Agricultural Dilemma*. Austin: University of Texas Press, 1990.

Zhang, Sarah. "America's Nuclear Waste Plan Is a Giant Mess: An Explosion Caused by Cat Litter at a Storage Site Was Just the Beginning." *Atlantic*, November 2, 2016.

Index

relocation, 24; and research, 72, 128; and rivers, 149, 160; safety, 124–25, 216n32, 217n44; sickness, 129–30; and Soviet Union, 91; and testing, 46, 49–50, 64; worldwide, 83

radioactive waste, 144–49, 151–59, 161–62, 165, 216n32

radioactivity: and agriculture, 115, 123, 126; and aquatic life, 19, 43, 56–57, 189n90; atmospheric, 75–76, 90; concerns over, 70, 106, 152; and contamination, 90, 146–47; control of, 84, 147–50, 154–55; effects of, 52, 74, 83, 147–50, 158, 162; and the environment, 145, 215n16; and fallout, 32, 75; local, 49; and nuclear waste, 151–52, 159, 163–64, 217n41; and rapid decay, 36; research, 62–63; sampling, 39–40, 43, 58, 73–74, 98, 102, 160; and sea water, 39–40; soil, 27; and strontium 90, 127

radioisotopes: and agriculture, 131–32, 135–38, 176n10; and Atomic Energy Commission, 212n91, 217n44; and crops, 127; and food, 9, 137–38; and medical research, 118; and nuclear waste disposal, 6; and pests, 124; and tracing, xii, 129

research funding, 47–48, 97, 122

rivers, 149–50, 152–53, 158–60, 163, 165

Roosevelt, Franklin Delano, 48

Rusk, Dean, 57–58

scientific knowledge: and Atomic Energy Commission, 149, 154; and atomic weapons, 43; and environmental science, 8, 22, 42, 58, 150, 155, 171–72; and fallout, 39, 80; and the natural world, 3, 167; and nuclear testing, 17; and policy-making, 6, 10, 154, 165; and US detection system, 93

Scripps Institute of Oceanography, 37, 151, 193–94n45

sea otters, 29–31, 43–44

seismology, 81, 87–88, 97–98, 102, 169, 198n3

Silent Spring, 170

Snake River, 160

Soviet Union (USSR): and Amchitka Island, 29; and disarmament, 80; and Francis Gary Powers, 103; and nuclear attack, 34, 84; and nuclear testing, 89, 101, 105–6; and nuclear test cessation, 43, 73, 78, 80, 85, 88–89, 91–92, 98, 102; and nuclear weapons, 127; and public opinion, 99; and seismic research, 104; and test detection, 29, 92; and the United States, 15, 35, 70, 84–86, 100, 103, 135

Strauss, Lewis, 35–40, 54, 58, 65, 70, 85–86, 187n65, 187nn68–70, 187n75, 187n79, 188n79

strontium 90 (Sr^{90}), 50–51, 62–67, 69, 73, 75, 127–28, 194n48, 210n57

Sutter, Paul, 5, 115, 177n16

Szasz, Ferenc M., 3, 14, 24, 180n2

Trinity site, 3, 13–15, 17, 40, 46, 48–49, 74, 167, 170, 189n3, 191n13

Truman, Harry S.: and agriculture, 115, 118; and the atomic bomb, 17; and atomic energy, 132; and Atomic Energy Act, 48; and Atomic Energy Commission, 7, 35, 94, 159; and environmentalism, 170; and fallout, 46, 49–51, 64, 75; and nuclear technologies, 168; and nuclear testing, 18–19, 22, 27; and nuclear waste, 144–45, 159; and public lands, 18; and the Soviet Union, 93

United Kingdom, 43, 89, 92–93, 104

United States Army, 34, 138–39, 162

United States Department of Agriculture, 126, 130, 195n61

United States Department of the Interior, 17, 19, 29–31, 132, 182n20

United States Forest Service, 1–2

United States Navy Department, 19, 38

United States Southwest, 18

United States State Department, 19, 100, 105